BEN SHE

Ben Shephard was born in 1948 and read history at Oxford University. He was a producer on the television series *The World at War* and *The Nuclear Age* and has made numerous documentaries for the BBC and Channel Four. He is the author of the critically acclaimed *A War of Nerves: Soldier and Psychiatrists, 1914–1994*, *After Daybreak: The Liberation of Belsen, 1945* and *The Long Road Home: The Aftermath of the Second World War*. He lives in Bristol.

BEN SHEPHARD

Headhunters

The Pioneers of Neuroscience

VINTAGE BOOKS
London

1 3 5 7 9 10 8 6 4 2

Vintage
20 Vauxhall Bridge Road,
London SW1V 2SA

Vintage is part of the Penguin Random House group of companies whose
addresses can be found at global.penguinrandomhouse.com

Copyright © Ben Shephard 2014

First published in Vintage in 2015
First published in hardback by The Bodley Head in 2014

www.vintage-books.co.uk

A CIP catalogue record for this book is available from the British Library

ISBN 9780099565734

Printed and bound in Great Britain by Clays Ltd, St Ives plc

Penguin Random House is committed to a sustainable future for our
business, our readers and our planet. This book is made from Forest
Stewardship Council® certified paper

MIX
Paper from
responsible sources
FSC   C018179

Contents

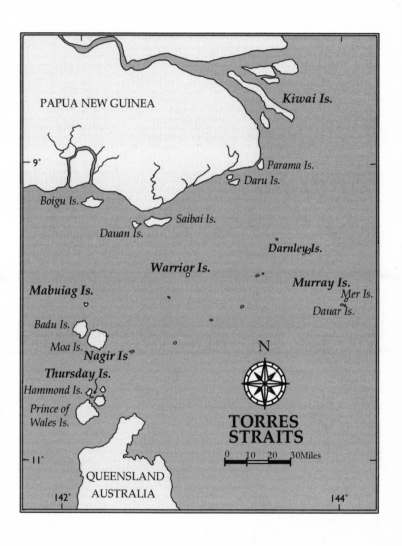

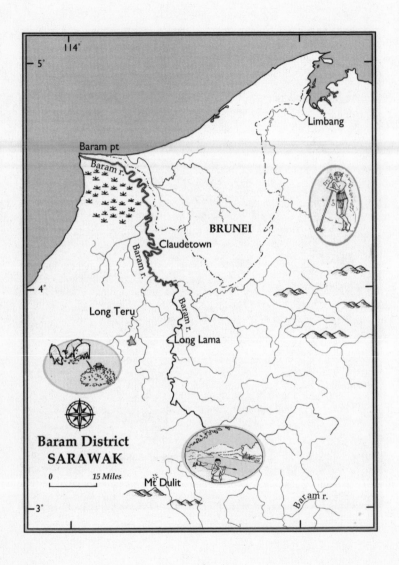

114°

5°

Baram pt

Baram r.

Limbang

BRUNEI

Claudetown

4°

Long Teru

Baram r.

Long Lama

Baram District
SARAWAK

0   15 Miles

Mt Dulit

Baram r.

3°

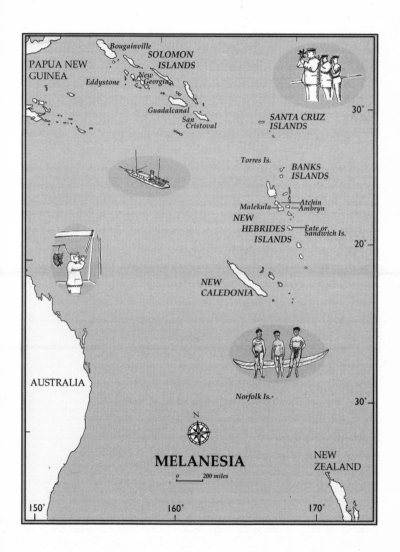

PAPUA NEW GUINEA

Bougainville

SOLOMON ISLANDS

Eddystone

New Georgia

Guadalcanal

San Cristoval

30°

SANTA CRUZ ISLANDS

Torres Is.

BANKS ISLANDS

Malekula

Atchin

Ambryn

NEW HEBRIDES ISLANDS

Eate or Sandwich Is.

20°

NEW CALEDONIA

AUSTRALIA

Norfolk Is.

30°

N

MELANESIA

0   200 miles

NEW ZEALAND

150°   160°   170°

Introduction

In the late nineteenth century, a generation after Charles Darwin published *On the Origin of Species* in 1859, glittering prospects seemed to be opening up for scientists in Britain. The theory of evolution was by now an accepted fact, but its implications for the different branches of science were still being explored and absorbed. One obvious task was to establish where man sat in the scheme of evolution, following the route mapped out by Thomas Henry Huxley in *Man's Place in Nature* in 1863, when he called for a Science of Man which would embrace anthropology, zoology and biology. The tools needed to do this were, it seemed, being rapidly developed.

One line of enquiry in particular had emerged. Clearly what set man apart from the other animals were his power of speech and the size of his brain. Anatomists found that in man the forebrain, the executive centre which controlled the senses and powers of decision, was enormous, whereas in fishes, amphibians and reptiles it was tiny. And the human cerebral cortex was four times the size of that of man's nearest relative, the chimpanzee.

How, then, did the human brain evolve? Why did it evolve as it did? In the 1870s, modern experimental neuroscience began, using electricity to stimulate the nervous system of animals and microscopes to observe the nerve cells of humans. Within two decades, researchers had established the location of functions within the brain, unravelled the way that the nervous system automatically governs the body's functions, and begun to discover how messages are sent between neurons and synapses. But these extraordinary advances only posed further questions – about human behaviour, man's relations to his fellow primates, and the human occupation of the earth. A generation of scientists went looking for the answers.

This book traces the intellectual journey of four of them, who met at Cambridge University in the 1890s and whose lives and careers were interlinked for the next three decades – William Rivers, Grafton Elliot Smith, Charles Myers and William McDougall. It starts with the neuroscience of their day and follows their voyages of discovery – both physical and intellectual – in peace and war. The journey takes in fields which today carry a variety of labels – neurology, psychology, psychiatry, anthropology, archaeology, to name only five – but which for these men were all part of the same enquiry. Rivers, for example, considered all his varied interests to be 'aspects of the same problem, the biological reaction of man to his environment'.[1]

The origins of this book go back a long way, to the summer of 1963 when, as a sad, solitary fourteen-year-old on holiday in Italy, I first read Siegfried Sassoon's book *Sherston's Progress*. I was unacquainted with the earlier volumes in Sassoon's trilogy, knew nothing about shell shock, and was bemused by the ironic and facetious tone of the writing. But one of the characters, W. H. R. Rivers – the doctor who treated Sassoon at 'Slateford Hospital' and became his father-confessor – stayed with me and became as much part of the furniture of the war as Haig or Hindenburg.

Two decades later, by now with a wife and two small children, I found myself one Easter in a small bungalow near Amiens rented by my well-meaning parents-in-law. The only thing to do was to drive across the flat, featureless plain to Albert and visit the battlefield of the Somme. It was an extraordinary experience to tread the ground where nearly 700 men of the Newfoundland Regiment had fallen on 1 July 1916: on my return to London, I wolfed down Sassoon's trilogy and the memoirs of Edmund Blunden and Robert Graves. By now I had got the Great War bug badly, but it was not until five years later, while working on a television series in Boston, that I got a clearer idea of what I could usefully do about it. After sampling several books on the psychological aftermath of the Vietnam War (with titles such as *Trauma and Transformation*), I conceived the idea of writing a history of military psychiatry, in order to establish whether the past really was as depicted by American psychotherapists. The resulting book, *A War of Nerves,* was finally published in 2000.[2]

Inevitably, my research quickly took me back to Rivers. But I soon discovered that, although he was the only shell shock doctor to be memorialised by a patient, he was far from being the only psychologist to have treated war-shocked soldiers. Countless doctors had been involved in the work and many had written books about it. In particular, Rivers' pupil Charles Myers had coined the term 'shell shock' while serving with the Army in France; another pupil, William McDougall, had written a wonderful account of the patients – ordinary soldiers, not poets serving in the Royal Welch Fusiliers – whom he treated back in England; and Rivers' friend Grafton Elliot Smith had published a wartime polemic on shell shock.

Furthermore, I learned that these men's work with shell-shocked soldiers was simply one chapter in careers devoted to a plethora of other intellectual projects: all were Fellows of the Royal Society. Rivers, Myers and McDougall had, for example, gone on a famous anthropological expedition in 1898, when for the first time the scientific methods of modern psychology had been applied to so-called primitive people. I soon became fascinated by the differences in temperament which began to emerge. Myers, I learned, was 'perhaps the ablest and most balanced mind among British psychologists of the twentieth century', but he seemed to be a pioneer by temperament: better at starting things than seeing them through to completion. I eventually managed to track down Myers' children, in deepest Gloucestershire. Brigadier Edmund Myers, a sprightly figure in his eighties who had only recently abandoned the hunting field, told me his father had 'lacked the common touch'; his sister, Joan, lent me the diary Charles Myers had kept in 1914 and 1915.[3]

McDougall's image was quite different. His life, one gathered, had been a tragedy: he was a genius who could have accomplished anything had he not been brought down by flaws of character – the exact nature of which was never spelled out. It was said that in his main book he 'seemed to do a great deal of packing for a journey on which he never starts'. My curiosity about McDougall grew when I attended a conference held at St John's College, Cambridge to mark the centenary of the 1898 expedition. Not one of the papers was about McDougall, even though he later became the leading British psychologist of his generation; his name was scarcely mentioned. He seemed to have been airbrushed from the Cambridge version of history, like an old Bolshevik in a photograph retouched in the Stalin era. What, I wondered, had happened on the expedition – and why did McDougall separate himself from his colleagues afterwards?

The big question about Rivers was why this 'wonderful' man, the 'great and good' Rivers (as Sassoon called him), should have chosen to associate himself intellectually with Grafton Elliot Smith, an Australian anatomist who dabbled in archaeology and anthropology. There seemed to be a 'tangible whiff of scandal' about Elliot Smith: the mere mention of his name reduced Cambridge anthropologists to apoplexy. My search for answers took me to Ashworth Special Hospital near Liverpool, today a special facility for the criminally insane but in 1915, as Maghull War Hospital, the setting for Rivers' and Elliot Smith's work with shell-shocked soldiers.[4]

However, I was soon faced with a quandary. In my book, I could either take the story of military psychiatry up to the end of the twentieth century, into the era of Post-Traumatic Stress Disorder, or I could write a detailed account of Rivers and his circle. I could not do both. After some debate, I opted to do the first in *War of Nerves*. But I remained fascinated by the intellectual world that the Cambridge group inhabited and eventually decided to return to it. *Headhunters*, which borrows its title from the account of the 1898 expedition which its leader, Alfred Haddon, published over a century ago – is the result.

My *Headhunters* is intended primarily as a character study, a group portrait of four men, not an academic survey of their scientific work. Inevitably, more material survives in some areas than in others. Myers' widow destroyed his papers (except those she thought would interest her soldier son); Rivers' papers were weeded after his death, and a personal diary he kept in Germany in 1892, which may have contained evidence as to his sexual orientation, has disappeared. Against that, however, I have made new discoveries which help to explain, for example, Rivers' chequered medical career in the 1890s, why McDougall got married in a hurry, and how Myers came to invent 'shell shock'. I am also probably the first scholar to make use of a confessional document which McDougall wrote on his deathbed.

I have also tried to give some idea of how science actually works: the passions, the irrational flashes, the moments of insight – the big ideas that work and the big ideas that are plain wrong. Churchill once said that history is always written by the victors. The history of science, too, tends to be Whiggish: usually written as a narrative of heroic progress towards the present, with most attention going to those who were proven right and little to those whose findings and theories did not ultimately stand up. But in the human sciences, this model simply doesn't work. Few psychologists are right about everything. They usually get some things right and others wrong, and our perspective is constantly altered as social and cultural changes take place around us. In the fairly recent past, for example, two projects which now seem completely insane secured solid institutional backing: one involved officials at America's top universities taking naked photographs of undergraduates on arrival, in order to decide who was fit to mate with whom; in the second, researchers domesticated, breast-fed and tried to teach language to a chimpanzee, in order to disprove the linguistic theories of Noam Chomsky. Who knows what present follies wait to be exposed by future historians? On the other hand, as the novelist Ian McEwan has argued, 'you never know when you might need an old idea. It could rise again one day to enhance a perspective the present cannot imagine.'[5]

Rivers and his friends were very much 'men of their time'; in some ways they seem light years distant from today's world. But they are also strangely modern, for the agenda they addressed is similar to that confronting the human sciences today. We are still faced with the questions these men grappled with. How far can the workings of the human mind be understood by science? And, in trying to reduce human nature to science, do we lose touch with what is distinctively human?

PART ONE

THE VOYAGE

I

The Alluring East

'A journal must begin somewhere,' Alfred Haddon wrote in his diary on Thursday 10 March 1898, 'and the start of a voyage is as good a place to begin with as any other.'

He was on board a ship making its way to the mouth of the Thames, pausing at Gravesend to take on several tons of explosives and to collect sailors left behind in London, pursued by innkeepers to whom they owed money.

It had been a long, tense day for Haddon. In the morning he and his family had taken a cab through the teeming streets of London's East End to the Royal Albert Docks, the new anchorage where large, modern ships were berthed, and made their way to the Dorsal Line's the *Duke of Westminster*. She was a 4,000-ton cargo boat bound for Brisbane in Northern Australia, carrying mails between Aden and Queensland, forty steerage passengers, 'excellently accommodated for about £12 a head', and eighteen second-class passengers. These included the Bishop of Queensland, his chaplain, and two other parsons in his wake; two miners from Cornwall, and a mining engineer; three ladies; and five members of the expedition to the Southern Ocean, sponsored by Cambridge University, which Haddon was leading.[1]

At about eleven, the crew had begun making noises about leaving, but it took a good half-hour before all the goodbye-sayers could be coaxed ashore. Finally, at about ten minutes past noon, the *Duke of Westminster* had cast off from the wharf, to the cheers of families and friends on the quayside. 'There was a protracted waving of handkerchiefs', Haddon noted. 'The tying of a number of handkerchiefs together and waving them up and down Auld Lang Syne fashion was duly appreciated and several times alluded to during the early part of the voyage.'

Haddon had not enjoyed the farewells. His conscience was not quite clear. He was leaving behind a wife and three young children, and going to the other side of the world in a venture by no means guaranteed of success. It had taken months of nervous strain to get the expedition going at all. No wonder he was impatient to get on with the voyage. 'We were a long time being piloted through a maze of large and small craft and the narrow entrances and of A[lbert] Docks – and at various spots friends of those on board met us again and again – with words of cheer & waving "hankies". At last we got clear, not sorry to be done with the painful leave-taking, & we steamed down the broad Thames glad to be at last on our road to the alluring East.'[2]

The ship was taking Haddon's team to the Torres Straits, which lie between Australia and New Guinea. Haddon had been there once before, and for years had dreamed of returning with a properly equipped expedition. Now, finally, he had realised that dream. But it had been a close-run thing.

Alfred Cort Haddon was forty-three years old, a tall, handsome man with a broad-boned, slightly simian face, and an intense physical presence. Vitality emanated from him – a simple, artless delight in the natural and human world. He was at once a dreamer and a man of ambition.[3]

Haddon was born onto the lower rung of the mid-Victorian middle class, coming from a long line of Nonconformist printers. His father ran the family business, but neglected it for religion and music, so that Alfred's childhood was spent in the grander suburbs of London and his boyhood in the humbler ones. His education was patchy and, mostly, Haddon educated himself: he was the archetypal bug-man. From an early age, his pockets bulged with animals and insects; he was constantly dissecting small creatures with his penknife. His youthful diaries, his biographer would later record, 'reveal the large (mainly Nonconformist) circle in which he grew up, his deep religious convictions, his wide interests, his physical energy and his entire devotion to natural science, whether botany, geology or zoology, especially the latter, almost all self-taught by reading and by observation. But books were too expensive to buy and difficult to borrow, so all the more dependence was placed on practical experiment, ducks and hens, rabbits and moles, rats and tortoises all assisting in the work.'[4]

By the time he was seventeen, Haddon was involved in running the family printing business. In the evenings, however, he wrote to famous men of science for advice on how to pursue his dreams. They did their best to put him off – no one could earn a living in science, they all told him, except perhaps in geology or medicine; far better to stay in publishing and devote his leisure to science. But Haddon persevered, took evening classes, and, after much cramming in Greek and Latin, was accepted by Christ's College, Cambridge in 1875. He was fortunate in his timing. Ten years earlier, the university would have been barred to Haddon: only in 1871 did it become possible for dissenters to take degrees there. And a decade before, physiology – the subject which Haddon chose to study when he had somehow scraped through the preliminary examination in Greek – had still been stagnating at Cambridge, caught in the old religio-anatomical mould, with research inhibited by rules against vivisection; not an experimental science, as it was in Europe. However, some at Cambridge had seen the need to change and in 1870 Trinity College had brought in Michael Foster, a young protégé of Thomas Henry Huxley, from University College London, to teach physiology. A Baptist turned agnostic, Foster was an inspiring leader and brilliant academic politician who revolutionised the teaching of science by his insistence on practical work and established a laboratory which by the end of the century was world-renowned.[5]

Haddon was at first frightened by Cambridge, where he was obliged to attend chapel and introduced to wine by his fellow undergraduates. But he quickly found his feet and his promise was soon spotted, not only by the sharp-eyed Michael Foster but by Foster's 'captain' – the great Huxley himself, the man who had created the template for a new breed, the professional scientist. By this time, Huxley's beginnings as a struggling apothecary were long behind him, and his popularising of Darwin a decade in the past. The radical, the man who had coined the word 'agnostic' and scandalised the public by casting scientific doubt on the Resurrection, had become the Pope of British science, at once scientific showman, sage, and conscience of the age. Although he never held a major academic chair and was forced to take on hack work to feed his family, Huxley was 'President of all the societies' (as Michael Foster called him) and had immense powers of patronage.[6]

Two decades after the publication of *On the Origin of Species* the main task of life scientists was to fill in the gaps in Darwin's scheme – and in Huxley's elaboration of it. One such area of interest was in the origins of the lowest forms of life, such as marine organisms: how they had developed and where they sat in the chain of evolution. This was the main focus of Haddon's zoological interest at Cambridge and in applying it, he acquired a technique of careful and minute observation, of the amassing of information, which he would carry with him through all his later scientific work.

Haddon did brilliantly in his tripos, distinguishing himself in comparative anatomy and zoology – on the strength of which he was able to spend January to June 1879 at the newly established Zoological Station at Naples. There he laid the foundations for his future career. 'His boyish hobbies with aquarium, insectarium and microscope had first interested him in lowly forms of life', wrote his biographer. 'Cambridge . . . widened and deepened those interests; Naples with its opportunities for original research gave the first impetus and for the next decade Haddon devoted himself to marine biology.' But how was he to advance? 'At the present moment', Michael Foster had written in 1874, 'the life of a scientific man is a very hard and unremunerative one. He can only get money by teaching or lecturing or writing and even then with such difficulty that very little time is left for original research.'[7]

Haddon's need to earn a living was all the greater because he was by now engaged to be married. His best friend at Cambridge was not a fellow scientist but John Holland Rose, a witty and talkative draper's son, then reading classics and later to become a distinguished historian. In May Week of 1877, Haddon and Rose had each paired off with the other's sister – Holland Rose with Haddon's lively and attractive sibling Laura; Alfred Haddon with Rose's quieter sister, Fanny. The arrangement, originally light-hearted, had stuck. John Holland Rose and Laura were married in July 1880; Alfred Haddon and Fanny announced their engagement a month later. After a period spent working as curator of the Zoological Museum and demonstrator in the anatomy department in Cambridge, the solution finally arrived: in December 1880, through Huxley's influence, Haddon was appointed professor of zoology at the Royal College of Science in Dublin. Nine months later, he and Fanny Rose were married.

In many ways life in Dublin suited Haddon. He was a devoted father to the three children that quickly appeared; Fanny proved an able and supportive companion, uncomplainingly accepting the privations of an academic wife's lot; and his colleagues included many fellow spirits, with whom he could argue and debate over tea and buns in the lunch hour. Haddon's gifts as an inspiring teacher emerged and he soon had a devoted circle of pupils.

An Irish friend later recalled Haddon at this time as 'a vigorous, restless, boisterous person, impatient, full of laughter, with a rapid, rather stuttering speech, the result of his impatience . . . His hair was jet black, worn long, with a large lock dangling over his sallow face, and he walked rapidly with an untidy gait and a slight stoop. If you met him in the street he would probably be dressed in an old velveteen jacket, trousers baggy at the knees and fringed at the bottom, and an old hat stuck on anyhow, and he would be posting along at about four and a half miles an hour. He rather liked shocking prudish people – who, according to modern standards, abounded in those days – and those whom he thought insincere or silly often thought him rude. People who knew him slightly found him an uncomfortable person but all who knew him well, loved him, he was so loyal and sincere. There were a good many people he just couldn't be bothered with, and I fancy he let them know it pretty plainly, but I don't think he had any enemies.'[8]

At Dublin, Haddon was able to continue his research into marine biology: in 1885 he organised the first deep-sea dredging expedition to the south-west coast of Ireland, which produced numerous new species of anemone and sea urchin. There were downsides, however. This was the period when Gladstone's government was trying to introduce Home Rule to Ireland and passions ran high on all sides. 'You have no idea what it means to be an alien in race and religion here', Fanny Haddon wrote to her mother. 'Even I am only just beginning to realise the intense scorn and hatred of the Irish towards all foreigners.' Although the Haddons felt some sympathy for the Irish cause, they were necessarily seen as Protestant foreigners and it soon became apparent that religious politics extended into academic life.

When Haddon failed to get a university fellowship in 1885, Fanny reported that 'it was the national, Catholic and presumably patriotic element that combined against Alfred'.[9]

By the end of the decade, Haddon was getting restless, feeling that he was 'perishing for want of research' and that 'the Department here is played out'. He began applying for jobs elsewhere, even going so far as to get his hair cut, but when he failed to get the professorship of biology in Melbourne he started to feel that it was hopeless. He could no longer go on 'retailing second-hand goods over the counter', he wrote, 'lecturing about coral reefs and tropical fauna which he had never seen'. He decided to go off to some marine biological hunting ground and do his own research. One day, on the cliffs at Howth, he told his wife of his plans. She agreed without a murmur.[10]

Where should he go? Haddon asked the great men of science. The naturalist Alfred Russel Wallace advised the West Indies, which was quite near and could offer the researcher a comfortable house with civilised surroundings. But Haddon was more attracted by the islands in the Torres Straits, the narrow strip of sea between Northern Australia and New Guinea, which T. H. Huxley had visited as the surgeon on the voyage of HMS *Rattlesnake* in the 1840s. Haddon found missionaries who had worked there and were willing to help him and, with Huxley's patronage, he was able to get a travel grant of £300. In July 1888 he set out to 'investigate the fauna, structure and mode of formation of the coral reefs' of the Torres Straits. A new phase in his life began.[11]

From the moment Haddon reached Thursday Island, at the western end of the straits, on 8 August 1888, it was the natives rather than the coral reefs that fascinated him. He had arrived at a time of huge cultural upheaval. By the early nineteenth century, the straits (named after Luis Vaez de Torres, the Spanish explorer who sailed through them in 1606) had become a well-frequented route for shipping, offering a shorter passage to Queensland and New South Wales. Yet it was also a dangerous one: the hundred or so islands the straits contained remained uncharted and when ships ran aground the islanders often killed their crews. Eventually, in the 1840s, this hazard to shipping led the Royal Navy to send three successive ships, the *Fly*, the *Bramble* and the *Rattlesnake*, to carry out hydrographic surveys of the waters. Unintentionally the navy had also commissioned the first anthropological accounts of the Torres Straits: the ships' naturalists described in some detail the islanders' great ocean-going canoes, equipped with double outriggers and rigging, which enabled them to travel great distances; and detailed their methods of warfare and religious customs. They also reported on their propensity for head-hunting.[12]

Much had changed when Haddon arrived a generation later: commerce and Christianity had come to the straits. The discovery that the pearl beds off the northern coast of Australia were the largest in the world and that the region was also rich in bêche-de-mer or sea cucumbers – much prized in Chinese cuisine as a delicacy and an aphrodisiac – brought a swarm of white men to the area, who employed the natives to wade for the shells in shallow water and later, when the beds close to shore were worked out, to dive for them a little further out. The 'pearlers' were rough, unscrupulous men out to make a quick profit. They forced the islanders to work for them and stole their women; fights with the islanders became common. In 1872 it was reported that the 'once confident and fearless' people of Mabuiag Island had become 'cowed and sullen, going bush'. In an attempt to control the exploitation, the whole of the Torres Straits Islands were annexed to the state of Queensland seven years later. More importantly, the London Missionary Society arrived in the islands in 1871. The missionaries were intent on protecting the islanders from the 'dark deeds' and 'infamous acts' of foreign seamen; but they also got rid of their traditional shrines, built churches, covered the nakedness of the women with muslin and calico, and attempted to impose their own puritanical moral code on the young men.[13]

On his arrival, Haddon was delighted to find that the Queensland government had put its officials at his disposal. He began his research by accompanying the acting Government Resident, Mr Milman, on a month-long voyage from Thursday Island to the northern islands, close to New Guinea. As Milman administered justice and sorted out disputes among the locals, Haddon submerged himself in their culture, examining how their houses were built (and often darting into them); bartering for masks, skulls, pipes, tomahawks and other curios; and persuading women to take off the missionary calicos and once more put on 'a scanty but sufficient grass petticoat'. Haddon's boyish, eager personality and physical presence gave him an easy rapport with the islanders. He had, his biographer records, 'the one thing needful for a field anthropologist, without which all work is barren, the gift for friendship'.[14]

He was also extremely nosey. 'I am finding some curious facts', he wrote in his journal. 'One point common to Toota and Prince of Wales Island at least, is that a girl may propose to a man! She sends to the chosen one a ring of string by the hands of a common friend or preferably of his pal or chum and through him makes an appointment in the bush.' On one of the islands he met a chief called Mamoose (who had his name engraved on a piece of brass round his neck, 'like the label for a bottle of wine') and was able to ply him with questions about the 'numerous forms of marriage custom designed to prevent intermarriage between certain families'.[15]

Haddon's greatest coup, however, was to persuade a group of old men on Prince of Wales Island to restage their traditional war dances. The resulting scene stirred him deeply: a tropical twilight with fires burning under a starry sky, waves lapping and drums rhythmically pounding. A wailing chant began and seventy warriors in full paint of red ochre, decorated with shell-ornaments and leaves, skipped from side to side, brandishing their weapons at their imaginary enemies. Hadden wrote in his journal:

> Altogether it was a vivid glimpse into savagery, all the more remarkable for the fact that half an hour's sail across the channel lands you into nineteenth-century civilisation! Hardly anywhere else in the world can the transition be effected so rapidly between the conventional manners of English society and the barbaric customs of the uncivilised savage. Years ago these islanders had a most unenviable reputation for ferocity, exhibiting a fierce determination to slay the encroaching white man – now they walk thirteen miles and dance the same day for the delectation of an Englishman they have never seen.

But, if the ways of the past could still be summoned up, most of the men on the islands now worked in the fishing industries and had begun to absorb European ways. On one occasion, Haddon observed George, one of his main informants, playing cards with his family. 'It was amusing to hear the English names of cards mixed up with native words, thus in the middle of to me unintelligible words would come

– "I have six tricks", "What are trumps-spades", "Give me a bloody heart", "that card will kill him"– and so forth.'[16]

After five weeks Haddon finally settled down to marine zoology, spending a month dredging the reefs off the island of Mabuiag, in the western end of the straits. He was excited to find living blue coral (*Heliopora coerulea*): 'I believe I am one of the very few zoologists who have seen the polyps expanded – if not the only one', he wrote in his journal. To get a closer look at the coral reefs, he also went down in a pearl-diver's suit, equipped with a heavy helmet, weights, and large boots, but he discovered little and found the whole experience unpleasant. When his suit started to leak he pulled the line and was winched to the surface. Far more agreeable were the evenings he spent 'yarning' with the local people, questioning the fishermen and policemen who were assisting him about life on the islands 'before white man he come – no missionary – no nothing'. 'We have very pleasant times together, laughing and talking', he noted. 'They are very ready to tell me all they can but too often the reply is: "Me young man (some 30 or so with a beard) me no savvy – old man he savvy".'[17]

This was one of the turning points of Haddon's career. He realised that not only were the old customs dying out, but the younger generation was forgetting them – and that, unless someone recorded them, they would be lost forever. When he reached Murray Island, in the east of the straits, in October 1888, and found that the old customs had survived better there, he began to conduct more systematic fieldwork, focusing especially on the islanders' dances and initiation rites. He also witnessed a funeral:

> As I saw the black silhouette of the canoe and its crew against the moonlit sky and sea, silently gliding like a veritable shadow of death, and heard the stillness of the air broken by the moaning of the bereaved ones, my mind wandered back thousands of years and called up ancient Egypt carrying its dead in boats across the sacred Nile – there with pomp, ceremony and imagery, here with simplicity, poverty and stern realism.

But, amid all the excitements, Haddon became increasingly frustrated by his own shortcomings. Pidgin English was 'a vile and

unsatisfactory medium for the communication of other than the
most elementary ideas' which made it very difficult to get 'connected
accounts of these old ceremonies'. 'To be a proper anthropologist,'
he wrote in his journal, 'one requires wider knowledge and more
versatile talents than I can lay claim to. He should be a linguist,
artist, musician, and have an extensive knowledge of natural and
mechanical science.'[18]

When Haddon got back to England in 1889, his mind was made up.
He wanted to be an anthropologist, not a zoologist; and he wanted
to live in Cambridge not Dublin. Although he wrote up his marine
research in several papers comparing the evolution of the sea anem-
ones of the Torres Straits with those in Ireland, that work was
completely overshadowed by the three huge monographs he quickly
produced on the languages, legends, myths and customs of the Torres
Straits islanders. He also gave numerous lectures to scientific gatherings and published popular articles on Papuan marriage customs and
dances in *Lippincott's Magazine*.

Haddon's writings had a considerable impact. At a time when it
was rare for British anthropologists to carry out extensive fieldwork
and most of the great names in the discipline operated from
armchairs, polishing up travellers' tales and missionaries' observations, Haddon had made a substantial new contribution. In July 1890
the Cambridge classicist J. G. Frazer, whose masterpiece of armchair
anthropology *The Golden Bough* had just come out, wrote to express
his admiration. Haddon's papers were priceless, Frazer told him,
'since the information they contain, if it had not been collected by
you, would probably have entirely vanished'. Such work would be
'remembered with gratitude long after the theories of the present
day (mine included) are forgotten, or remembered only to be despised
as obsolete and inadequate'. Such praise was gratifying. But, although
Haddon took the bold step of moving his family from Dublin to
Cambridge in 1893, his efforts to switch careers and to find a secure
academic perch in England were not successful. He was forced to
take Huxley's advice – 'don't burn your ships in a hurry' – and to
remain in Dublin himself.[19]

So he made the best of it. And in fact, Ireland offered him a rich
harvest. At this time anthropologists were interested in delineating

different physical types, and had recently developed more sophisticated instruments to carry out their research. The technique known
as 'anthropometry', or 'the Art of Measuring Man', was designed
'to obtain as exact a record as possible of the structure and functions of the human body, with a view to determining . . . how far
these are dependent on inherited or racial factors and how far they
vary with his environment'. Haddon, with his training in anatomy,
took up physical anthropology with enthusiasm and quickly
mastered its techniques. He also fell under the sway of the poly-
mathic Francis Galton, a cousin of Charles Darwin and a powerful
figure in Victorian science. Obsessed by the role of heredity in
producing men of genius (they were all men), Galton had become
interested in finding ways to assess intelligence and in 1884 had set
up an anthropometric laboratory at the International Health
Exhibition in London, which measured volunteers' physical strength
and sensitivity to stimuli. Haddon was further influenced by John
Beddoe, a Bristol doctor who had carried out a survey of the British
population, using traits such as hair and eye colour to assess racial
origins, and by two French anthropologists, Paul Topinard and René
Collignon, who had similarly made detailed regional surveys of the
French population.[20]

'The physical anthropology of Ireland is almost an untrodden field,'
Haddon declared in 1892, 'yet there is no part of the United Kingdom
which promises a richer harvest for the investigator.' By then he had
opened up an Anthropometric Laboratory in Dublin, modelled on
Galton's in London, where the techniques of human measurement
were practised on student volunteers, and in the long vacations he
was pitching his tent in areas, such as the Aran islands off the Atlantic
coast of Galway, which might help 'unravel the tangled skein of the
so-called "Irish Race"'.[21]

Finally, in 1894, he secured a toehold in Cambridge by taking over
the lectures in physical anthropology which were now included in the
human anatomy syllabus at the university. The following year (while
continuing to teach biology in Dublin) he offered Cambridge students
lectures in ethnology and sociology and recruited them to his new
anthropometric research project. Then, after years of busy networking
on committees, Haddon achieved a real coup. All over Europe,
academics and intellectuals were concerned that the arrival of the

machine age and the flight of population to the cities were wiping
out old customs and cultures: they must be preserved before they
disappeared altogether. Haddon persuaded the British Association for
the Advancement of Science to set up an ethnographic survey of the
British Isles – a huge racial map of these islands, focusing on isolated
populations and their ancient customs, to be compiled at the local
level by amateur historians and archaeologists using the techniques
of physical anthropology.[22]

Haddon himself plunged in enthusiastically. In July 1896, Ernest
Rutherford, a young physicist from New Zealand only recently arrived
in Cambridge, reported that he had gone with Haddon, armed with
cameras, calipers and record cards, to the village of Barrington, just
outside Cambridge, to 'get the mean average of the East Anglian type',
by measuring the skulls of the men and photographing the games of
the children. Rutherford was unimpressed, however. 'You can't imagine
how slow-moving, slow-thinking the English villager is', he wrote
home. 'He is very different to anything one gets hold of in the colonies.' But for all Haddon's efforts, the scheme was soon in trouble. It
became clear that it was overambitious to expect amateurs at a local
level to carry out so complex a task. Moreover, as he was passionately
engaged with the movement to save 'vanishing knowledge', Haddon
well knew that the need to preserve the ways of the past was greater
in Oceania than in Britain – the Tasmanians had already vanished
without a trace.[23]

If 'the great need of anthropology just now [was] fieldwork among
savage people', as Haddon wrote in Nature in 1897, how should it be
conducted? In his book Evolution in Art (1895) he had tried to use the
principles of evolutionary biology to understand the development
of art forms in primitive people – and produced an intellectually
futile exercise. It was pointless, he now realised, to trace out the
history of a pattern or design, as if a human artefact were a shell
or anemone: you had to understand the thinking – the psychological
processes – which lay behind it. So he turned to a discipline which
had only recently started to be taught at Cambridge. If experimental
and observational psychology could be used in the field, Haddon
believed, it might be possible not only to understand 'native actions
. . . from a native and not an European point of view' but also to

produce results of scientific objectivity, firmly grounded in experiment.[24]

By the middle of the decade Haddon had decided that the only
way to give his career a new direction was by realising his long-held
dream of returning to the Torres Straits with a full-blown expedition.
He was beginning to look for the men and the money to make it
happen.

2

Picking the Team

In January 1897, Haddon told his friend Patrick Geddes of his plans for a second expedition to the Torres Straits. He intended to take a team of six, to include 'a linguistic expert, a trained experimental psychologist and one or two young men to train as field anthropologists'. His conception of the project had broadened considerably. He now believed that the 'psychology and sociology of savage peoples' were 'really of more importance' than their 'physical characteristics'. 'I am certain', he told Geddes, 'that the best way to start a School of Anthropology at Cambridge is to boldly go in for research work of this kind.' By then Haddon had secured some institutional backing. The Expedition would carry the name of Cambridge University; a committee of supporters, on which both J. G. Frazer and Michael Foster sat, had been established; and some money had been promised by the university. It was time for Haddon to start assembling his team.[1]

He already had a linguist lined up: Sidney Herbert Ray, a teacher with a remarkable gift for languages. Now forty years old, Ray had grown up in East London and left school at fourteen to work in a City office before training as a teacher. In 1882 he was appointed as assistant master at a school in Bethnal Green, a post he was to hold for the next forty-one years. Ray's lifelong interest in languages led him to work on Sanskrit and African languages, but he made little real headway until he happened to pick up a second-hand copy of a famous beachcombing book, the British sailor William Mariner's account of his experiences on the Pacific island of Tonga. This fired Ray with enthusiasm for the Pacific Islands and, encouraged by a retired missionary who had worked in Melanesia, he quickly established himself as a recognised expert on the languages of Oceania, without ever leaving Bethnal Green. He had already helped Alfred

Haddon to write up the material on language from his first Torres Straits expedition, though the extensive word lists Haddon had collected proved insufficient for Ray to make a proper study of grammatical forms. Now Ray was the obvious candidate for the second expedition, but it took Haddon some time to get the London School Board's agreement to release the teacher and he had to pay Mrs Ray £70 to compensate her for the loss of her husband's earnings.[2]

Soon afterwards Haddon engaged as the expedition's photographer Anthony Wilkin, an undergraduate at King's College, Cambridge, and the precocious author of *On the Nile with a Camera*, a racy account of his travels in Egypt. A jovial, outdoor-loving, Old Harrovian extrovert, Wilkin would certainly enliven any group. Unfortunately, however, two of his tutors at King's also wanted to join the expedition. The historian Oscar Browning offered himself to Haddon as a representative 'from the political side of anthropology', while the political philosopher Goldsworthy Lowes Dickinson admitted with his usual diffidence that he was an 'amateur, with no special knowledge of anthropology' but was 'interested in some of its problems'. Neither of these Cambridge figures was remotely suitable for an expedition to the tropics. 'OB' was sixty years old, a corpulent Old Etonian snob, and 'a hearty trencherman who ate lobster for breakfast' and delighted in corrupting handsome young men. 'Goldie', future mentor of E. M. Forster and Leonard Woolf, was a gentle aesthete, masochist, and boot fetishist, famously absent-minded and unable to cope with everyday life. Haddon gently declined their services, though it took a second letter before Lowes Dickinson got the message.[3]

The obvious candidate for the psychological role was William Rivers, the well-respected lecturer in psychology at Cambridge. But Rivers was not interested. So Haddon turned instead to Rivers' most brilliant pupil, who was also a protégé of his own – a young man called Charles Myers.

Then twenty-four, Myers was the son of a wealthy clothing merchant whose family had risen from a pawnshop in Chelmsford in Essex, to become bastions of the West London Jewish community and seat holders in the Bayswater Synagogue. Charles's mother came from another clothing dynasty which had by the late nineteenth century diversified into more gentlemanly pursuits such as music and archaeology. The milieu in which young Charlie grew up – in Bayswater and Notting Hill – was bustling and bourgeois, intellectual and musical; as a young man he would rise at 6 a.m. to practise the violin. Myers excelled at both classics and science at the City of London School, in the shadow of St Paul's Cathedral, and, with his mother's backing, was able to resist paternal pressure to enter the family business. Instead he went up to Cambridge to study medicine and to organise a chamber-music group in which he played the violin. He specialised in physiology but was excited by Alfred Haddon's lectures on anthropology and William Rivers' on psychology.[4]

Myers was clever, but not a man of ideas; his interest was in people. 'It was the men whom he met even more than what they taught him that attracted him,' one of his pupils later remembered, 'and it is very clear that it was their character, at least as much as their scholarship which carried him away.' 'Fired by my teacher A. C. Haddon's enthusiasm, I became passionately interested in physical and racial anthropology', Myers himself recalled; and when Rivers began to lecture on the senses and on experimental psychology, 'holding a practical course in both subjects in a small room lent to him by Foster in the physiology laboratory', he was further attracted.[5]

When Haddon asked him to go to the Torres Straits, Charles Myers was living in his parents' house in Holland Park while finishing off the clinical part of his medical degree at St Bartholomew's Hospital. He was an excellent musician, but his experience of psychology was limited, his only publication to date being an anatomical analysis of skulls found in a Suffolk churchyard. This was obviously on his mind when, soon afterwards, he ran into William McDougall, another former pupil of Rivers, now working as a doctor at St Thomas's Hospital in London. Myers wrote at once to Haddon:

> McDougall, whose existence I had forgotten . . . is the ideal man . . . in every way for your expedition. I may be wrong, but he is in my opinion the most brilliant and the most promising man who has appeared here in recent years. And I think my views will be confirmed by such as know him.

The timing was good, too. McDougall's current job as a house physician at St Thomas's would end in February, leaving him free to join

the expedition and he was willing to stay on in the southern hemisphere after October if he found the work interesting. Myers confessed
that he had 'inwardly felt all along that I could not profitably pursue
psychological investigations unaided & in this Rivers quite agrees with
me'. As he wrote to Haddon:

> McDougall is especially keen on psychology. *Entre nous*, he is writing
> a thesis for a St John's Coll. Fellowship on a purely psychological subject.
> Rivers would have suggested him had he thought there was a probability
> of his accepting. But the man appears to be intensely keen & would
> (only if essential) pay part of the expenses.
>
> Over and above all, he is an exceptionally good sort – a thorough
> sportsman, strong and hearty – and indeed I went so far as to say that
> if he would go, I would go too . . .
>
> What say you?

A day later, Haddon had a letter from McDougall himself. He at once
invited him to join the expedition.[6]

Like Charles Myers, William McDougall (then aged twenty-seven)
came from a nouveau-riche background, but in his case the fortune
derived from self-raising flour. In 1864 the four sons of a Manchester
teacher turned industrial chemist, operating in business as the
McDougall brothers, developed and produced a patent substitute for
yeast, thus initiating a revolution in home baking and establishing
McDougalls as a household name for self-raising flour. One of the
brothers, Isaac Shimwell McDougall, despised this business and spent
most of his life vainly trying to break away from it. 'Although he
became chiefly responsible for the chemical business, which brought
in large profits,' his son William later recorded, 'he built up alongside
it an iron foundry in which to manufacture his own mechanical inventions, and later a paper-pulp factory for the same purpose. And in
these enterprises he spent a large part of his profits.'[7]

It was the same with religion: Isaac's 'active mind with a streak of
originality' led him to belong to a succession of Christian sects, without
in the end staying with any of them. His son remembered him as
'benevolent and affectionate, with a strong taste for poetry and music;
as in his business, so also in religion, art and domestic affairs, he was

masterful, erratic, unpredictable and always naïve'. McDougall remembered his mother as 'a strikingly beautiful example of the fair, calm,
introverted Nordic'. He believed himself to be a blend of Highlander
and Saxon – of the Mediterranean and Nordic races which had
produced the English people.[8]

One of five children, William McDougall was brought up in a
comfortable and intellectually stimulating background, in a large house
with a tennis court, paddock and stabling, in the Manchester suburb
of Broughton. A precocious child who excelled at all branches of the
curriculum, he left school at fourteen and was then sent off to Weimar,
because his nonconformist father regarded English public schools and
universities as 'strongholds of Toryism'. McDougall entered Owen's
College Manchester at the age of fifteen and, three years later, his
father having relented, was admitted to St John's College, Cambridge,
to read medicine.

It was there that McDougall's lifelong ambiguity towards groups
first revealed itself. 'I was a freshman just turned nineteen', he wrote
later:

> My fellow freshmen were for the most part fresh from school, while I
> had graduated with first-class honours from a provincial university.
> While they had the childish outlook of the average public-school boy,
> I was in many ways extraordinarily mature. The result was that I lived
> a double life. As a freshman I took part in and enjoyed all the many
> boyish activities that make the daily round of the average undergrad-
> uate; I joined all the clubs and rowed in the college boat; I wined and
> sang and played cards. At the same time I looked on critically, despising
> these pursuits a little as somewhat childish.

In the pictures that survive of the rowing, tennis and athletic teams
McDougall joined at Cambridge, he invariably stands somewhat apart
from the flannelled fools and lounging hearties, with a semi-sneer on
his features. Looking back, he wrote that 'I have never fitted neatly into
any social group, never been able to find myself wholly at one with
any party or any system; and, though not insensible to the attractions
of group-life, group-feeling and thinking, have always stood outside,
critical and ill-content. I have participated in the life of many groups,
scientific, medical, academic, and social, but have belonged to none.'[9]

Like Myers, McDougall was dazzled by the excellence of the physiology department which was 'then in its prime at Cambridge and full of promise of infinite progress'. Most notably, Michael Foster's protégés, Walter Gaskell and John Langley, were uncovering the mysteries of the autonomic nervous system, the way the body's nervous system automatically regulates functions such as digestion, heart rate and temperature. McDougall also attended Haddon's and Rivers' lectures. Intensely ambitious, he then used his years of clinical medicine at St Thomas's in London, to carry out research in the hospital's physiological laboratory, where Charles Sherrington, a brilliant researcher from Cambridge, was working on the 'feed-back' mechanism in muscles. McDougall was very attracted to what then seemed to be the most exciting area of medical research.

But there was another side to him. While at Cambridge he had read William James's masterpiece *The Principles of Psychology*, published in 1890, and been beguiled by the prospect of reconciling the physiology of the brain with the understanding of the mind:

> I had while still an undergraduate, determined that a life devoted to the study of the nervous system was the most desirable of all; for in the brain, it seemed to me, were locked the secrets of human nature. But James showed me that neurological research is not the only road to the uncovering of these secrets, and led me to believe that they should be approached from two sides, from below upwards by way of physiology and neurology, and from above downwards by way of psychology, philosophy, and the various human sciences.

As part of this bold plan – and because he was an intellectual show-off – McDougall wrote two separate theses when he applied for a fellowship at St John's in 1897: one, a piece of solid research on the problem of muscular contraction; the other, a brash sketch of a programme for reconciling the physical and the psychological approaches to the brain and the mind.[10] In November 1897, while preparing to go to the Torres Straits, he heard that he had been awarded the fellowship.

There was no denying McDougall's brilliance. Although only three years older than Myers, he was intellectually much more mature, much more his own man. While admiring both Haddon and Rivers he had been much less exposed to them than Myers. McDougall was

handsome, athletic – and very arrogant; but there was also an inse-
curity about him. He had turned his back on his Manchester roots,
become ashamed of the grimy factories from which his family's wealth
came, lost his religious faith, and his mother. While at St Thomas's
he had also suffered an emotional upset – probably a love affair gone
wrong. Foreign adventure offered an escape.

Soon after McDougall agreed to come to the Torres Straits, things
began to get complicated. In August 1897, seven months before the
expedition was due to sail, Haddon received a letter from a colonial
administrator in a remote part of Borneo, who had read about the
expedition in the press. Charles Hose (a Cambridge man himself)
suggested that Haddon's team, as well as visiting the straits, should
come to his area, where he could provide unique opportunities to
observe primitive people: 'In Baram is still to be seen what cannot be
seen in other parts, the people as they were hundreds of years ago,
as regards their customs, but obedient to my government.' Haddon
at once decided to widen the scope of the venture, though it would
mean having to find more money. He also consulted his team. William
McDougall responded enthusiastically and volunteered to go to
Borneo too but Charles Myers was more cautious. Borneo would
clearly be 'an ideal hunting ground for anthropologists; probably
affording better results than the straits', but going there would upset
his arrangements for returning to Bart's Hospital and would depend
on 'whether physical and mental health permit'.[11]

The expedition's enlarged scope made it doubly important that it
contain a zoologist. But it proved difficult to find one. Over the summer
Haddon approached several well-respected scientists. Some expressed
interest but ultimately none of them proved willing to put his career
and finances on hold for a year. So Haddon was forced to consider
alternatives in Cambridge. One possibility was a recent graduate who
had little experience in collecting data but had done 'a considerable
amount of shooting and some entomology', but McDougall's brother
knew him and, when sounded out, reported that 'his mind was a very
cramped and rather peevish and childish one . . . I think he would be
rather a bore.' Fortunately, his family forbade him going.[12]

Then, in October 1897 another candidate emerged.

Charles Seligman was born in London on Christmas Eve 1873, the only child of a rich Sephardic Jewish wine merchant. At an early age, he showed an interest in natural history, collecting butterflies and carrying out chemical experiments. He was sent to St Paul's, the London day school, but did not enjoy it. Myers wrote of him:

> Lonely and unhappy at home, reserved and discontented at school, he would often play truant to satisfy his growing interests in animal and plant life, spending his time collecting, dissecting and reading. His mother, an invalid, would sometimes remove him from St Paul's to spend a term with her at a seaside resort. On these occasions he educated himself by reading widely in the local public library. When he was about sixteen he lost his father, and his mother died not long after. On her death, an uncle, his guardian, arranged for him to be housed in a family of relatives between whom and himself unfortunately there existed little affection or sympathy. He formed friendships with far older men who encouraged him in his tastes.

One of Seligman's mentors was Frederic Michael Halford, a Jewish cloth merchant and amateur naturalist, who would later become Britain's leading authority on dry-fly fishing, writing for the *Field* under the pseudonym 'Detached Badger'. It was in Halford's Sunday afternoon salons that he met a young man of roughly his own age, Charles Myers.[13]

Once Seligman had managed to get into St Thomas's Hospital, his ability began to show itself. He was not interested in medicine itself but in research. While still a student he began to study tropical diseases, bodily abnormalities in fish, and the heredity of hair and eye colour.

Seligman was eager to join the expedition. He told Haddon he was already planning to be in Australasia the following year, would pay his own fare, and was prepared either to be a full member of the expedition or to be loosely affiliated to it. But Haddon was not keen to have another doctor, and Jewish to boot, on board, so he took soundings. McDougall reported that Seligman was 'a pleasant man to get on with, certainly very keen on science generally and a man of some ability'. Seligman's chief at St Thomas's had 'never come across a man who is more keenly interested in natural science', albeit 'wanting dexterity in his fingers' and not 'a skilful dissector or manipulator'. Haddon even asked Myers' mother about Seligman. 'I quite admit

that his appearance is not exactly prepossessing', she replied, referring
to his Jewish looks, but thought him amiable and friendly. Charles
Myers, who had known him 'a long time since we were quite boys',
conceded that 'his looks are certainly unattractive' but said he was
'never in the least sulky'. They had spent six months together in
Norway without a cross word.

Haddon still hesitated. He asked McDougall again, who replied: 'If
it is a question of Seligman or no one I think he would be a desirable
addition to our party.' In the end what decided it was that no zoologist would come unless he was paid, whereas Seligman had independent means and was willing to pay his own way. So he would have
to do. Anyway, Haddon found that Seligman improved on closer
acquaintance. His job at St Thomas's having recently come to an end,
Seligman immediately began preparing full-time for the expedition.
He addressed the practical questions, such as the right drugs and food
to take with them, bombarded Haddon with detailed enquiries – 'I
suppose fish hooks and needles are the mainstay?' – and went to
Cambridge to see experts recommended by Haddon, who was by now
lecturing in Dublin.[14]

But then Haddon's preparations were pulled even further out of
shape. Late in November 1897, William Rivers changed his mind and
asked if *he* could come too. 'My plans have been disturbed by the
recent death of my mother and by the fact that I have been rather
run-down [and] feel very much in want of a holiday', Rivers told
Haddon. 'If you will have me I would like to join your expedition.'[15]

3

'I should go in for insanity'

This was not the first time William Rivers had gone to sea for the sake of his health. A decade earlier, as a newly qualified doctor, he had taken a job as ship's surgeon and sailed to Japan and North America, characteristically spending much of the time reading the works of Herbert Spencer, the Victorian sociologist and polymath. Ill health, sea voyages and restless intellectual curiosity had defined Rivers' life to date.[1]

William Halse Rivers Rivers (*sic*) came from a Kentish family and was born on 12 March 1864 at Constitution Hill, Chatham, just a stone's throw from the great naval dockyard. His grandfather was a naval officer, and his father, Henry Rivers, after taking a degree at Trinity College, Cambridge, became an Anglican vicar who held a succession of posts in the Medway area. The other side of the family was more exotic. His mother, Elizabeth, was the daughter of Thomas Hunt, a Dorset farmer who had developed a career as a speech therapist, using breathing exercises, muscle control and the building of a patient's confidence to cure stammering. Elizabeth's brother, James Hunt, a bold and resourceful man, trained in his father's techniques, published several books on speech disorders and, armed with a doctorate bought from a German university, set up in practice as a speech therapist in Regent Street. In the summer months he ran a residential clinic at the seaside, first in Swanage and then, assisted by his vicar brother-in-law, at Hastings in Kent. He also became involved with the new science of anthropology.[2]

When Hunt died prematurely in 1869 – worn out by the strain of sustaining two careers – Henry Rivers produced a seventh, 'greatly enlarged and entirely revised', edition of his late brother-in-law's *Stammering and Stuttering*, and took over his speech-therapy practice, soon moving to Knowles Bank, a large house enfolded in woodland near the fashionable spa town of Tunbridge Wells. His patients included some late-Victorian celebrities. In April 1873, for example, the Oxford mathematician, the Rev. Charles Dodgson, whose speech hesitation made reading aloud to congregations a particular trial, wrote asking for help with the letter 'p', and came down to Kent several times for treatment. There he played with the Rivers children – Willy, Charlie, Ethel and Katie. According to Katie's later memories, the author of *Alice in Wonderland* preferred the girls to the boys.[3]

Most of Henry Rivers' patients were teenage boys, however, who boarded with the family. One of Will's earliest memories was of taking part in a debate about 'monkeys' with his father's pupils. Perhaps not surprisingly, Will Rivers stammered badly. We don't know whether this condition was inherited or acquired by imitation of his home surroundings, maybe even a bid for his father's attention; or whether Will was treated on the Hunt system. In his later years Rivers remarked more than once that the best thing to do about a stammer was to 'forget it'. His own seems to have diminished through the years, but he was never completely free of it. He once told the writer Arnold Bennett that 'the affection is due to a defect in the brain which gives contradictory orders simultaneously when disturbed in a certain way'.[4]

Speech impediment or not, Will Rivers was clever. In 1877, when he was thirteen, the family moved again, to a dank suburban house, so that he could attend Tonbridge School as a day boy. There is no evidence that he suffered any of the miseries undergone by E. M. Forster at the school a decade later: though sensitive, intelligent, and frail, William Rivers was not perceived as an outsider. He ran in the cross-country race and was elected a member of the school debating society. At the age of sixteen, however, the first disaster of his life took place: an attack of typhoid fever forced him to miss the last year at school and to abandon his dream of winning a scholarship to Cambridge. For the rest of his life his health was fragile. 'He was not a strong man and was often obliged to take a few days' rest in bed and subsist on a milk diet', one friend later recalled.[5]

Rivers spent a year recuperating amid Kentish woods and hop fields, and then, following the suggestion of a military surgeon who had worked with his father, went to Bart's Hospital in London; he now intended to join the Army Medical Corps. Bart's was London's oldest medical school, in the heart of the City next to Smithfield Market and the Old Bailey, serving a very diverse population, and renowned for the excellence of its teaching. Rivers flourished in this more demanding environment, though it took a sea cruise in the Pacific in 1886 to get his health back after he had qualified at the early age of twenty-two. After a brief spell at Chichester, he returned to Bart's as house physician, where he was able to undertake some research on nervous illness. He was now seen as a young doctor of exceptional promise: 'a gentleman of high character and great professional attainments, a laborious and intelligent observer, kindly and judicious in his management of patients, and in every way useful and trustworthy as a Resident Medical Officer'. Armed with glowing testimonials from his chief at Bart's, Rivers arrived at the National Hospital for the Paralysed and Epileptic, in Queen Square, Bloomsbury, as junior house physician in January 1890. He was twenty-six.

Rivers had come to the medical equivalent of Mount Olympus. In the 1890s, neurology – what we now call 'neuroscience' – was the most exciting and stimulating area of medical research and the National Hospital (always known as 'Queen Square') an international centre of excellence. Over the previous decades, experiments on animals (using electricity to stimulate their brains) and observation of men wounded in war had given doctors a much more precise understanding of the localisation of functions in the brain; this had made it possible for the first time to perform surgery on that organ. At the same time, medical theorists had begun to speculate about the place of the brain in evolution and its relationship to the nervous system.[6]

At Queen Square, the young doctor was able to accompany on his ward rounds a great figure such as David Ferrier, a dapper Aberdonian with a trim moustache and jaunty bedside manner, who had plotted the geography of the brain in his animal experiments; listen to the speculations on the evolution of the human brain of the dry, melancholy Yorkshireman, John Hughlings Jackson, whose years of patient observation had transformed the understanding of nervous disorders such as epilepsy; and assist the brilliant young surgeon Victor Horsley in his pioneering operations and research into the nature of electrical currents in the mammalian brain.[7]

Rivers also made friends with a fellow house physician. Henry Head was three years older, came from a wealthy North London Quaker family, and had been to Charterhouse and Cambridge. He was loud, bumptious, and opinionated, held advanced views on social questions; and had just returned from two years of medical research in Prague, where he had studied with the physiologist Ewald Hering. He irritated his Queen Square colleagues by constantly referring to research in Cambridge and Germany but Rivers was excited by his enthusiastic accounts of his time abroad.[8]

For all the intellectual stimulus it provided, Rivers' position at Queen Square was a punishing treadmill, especially when he was promoted to registrar in November 1890. Founded in 1854 in a couple of houses on the north side of Queen Square, the National Hospital was the only specialist hospital of its kind in Britain and every day a stream of patients came to its doors with epilepsy, sclerosis, brain tumours, Parkinson's disease and paralysis, and the registrar had to take the clinical history of each of them on admission. He also had to visit all the patients twice a day; attend the consultant on his ward rounds and supervise treatments; organise the removal of patients showing mania, imbecility or feeble-mindedness; make sure all the equipment was in good order; conduct post-mortems; dispense drugs when the dispenser was away; prescribe for out-patients; monitor the food and ventilation; and not leave the hospital except by permission.[9]

Eventually, Rivers could take no more. After suffering from nervous exhaustion for several months, he was forced to resign from the National Hospital in May 1892. Of all the blows he suffered in his life, this was probably the hardest to bear, a humiliating failure which closed the doors of neurology. The registrar's job was meant to be an ordeal, designed to test a young doctor's resilience – and Rivers had failed it. Where then would his future lie?[10]

He travelled to Germany and spent several months in the town of Jena, twenty miles from Weimar, learning the language, visiting art galleries and attending lectures at the university. He probably went at the urging of Henry Head, but a spell in Germany was a standard part of a British medical education in those days. Although contemporary guide books dismissed Jena as 'a dull, dingy, antiquated town, with a large lunatic asylum conspicuously situated on a hill above', Rivers found the atmosphere congenial. He was particularly stimulated by the lectures of the German idealist philosopher Rudolf Eucken and two well-respected psychiatrists, Otto Binswanger and Theodor Ziehen (best remembered today for treating the demented philosopher Friedrich Nietzsche), who happened to be authorities on the disorder he was himself suffering from – neurasthenia, or nervous exhaustion. Fired up by these great men, he decided to change tack. 'I have come to the conclusion that I should go in for insanity', he wrote in his diary.[11]

On his return to England, Rivers felt ready to take a new step. He gave a paper on neurasthenia at his old hospital, telling his audience that the disease was 'more common than supposed', and prescribing 'a complete holiday' as the best treatment. Drugs, he said, were 'worse than useless'. Clearly he spoke from experience. Then, in November 1892 Rivers was taken on as a clinical assistant at London's oldest mental institution, Bethlem Royal Hospital, then still in Lambeth, in the building which now houses the Imperial War Museum. The former 'Bedlam' admitted 'all persons of unsound mind, presumed to be curable' for 'maintenance and medical treatment', but tried not to take 'pauper lunatics' and 'incurable patients' who would be sent to the new county asylums. Clinical assistants, usually medical-school graduates looking to specialise in the treatment of mental health, were employed on six- to twelve-month contracts to assist the physician on ward rounds, write notes on patients on admission and then keep progress reports.[12]

Rivers was at once plunged into the realities of Victorian lunacy. In his first two months he admitted several women with religious mania ('the devil was inside her'), men suffering from alcoholic delusions, tradesmen whose business worries had triggered breakdowns, depressed teenagers, and a tea-planter from India with 'General Paralysis of the Insane', soon to be recognised as a symptom of tertiary syphilis. Rivers' case notes, in his tiny but then still legible hand, show him patiently exploring the patients' social circumstances – the clerk in holy orders who had lost all his money, the adolescent worried about stealing from his siblings, the father of eleven children now unable to provide for them. The records at Bethlem are also a reminder of how crude psychiatry at this period still was. The diagnosis, by the medical superintendent, R. Percy Smith, fell under a few simple headings: melancholia; delusions; alcoholic mania; general paralysis of the insane. The drug treatments were equally primitive, consisting usually of the bromide derivative, hyoscine, used to sedate violent or insomniac patients.

One of Rivers' cases was a 24-year-old married woman admitted to Bethlem suffering from what her GP called 'excitement of housekeeping after confinement' – post-natal depression in modern parlance. 'She has since birth of child', Rivers wrote, 'always felt that something was wrong but did not know what, has been very restless and not able to settle anything; has not in any way lost affection for her husband and children, though occasionally has felt her children would be better dead – has not thought of harming them – has thought she would be much better dead and has at times thought of suicide.' She was discharged five months later, in line with Bethlem's fifty per cent recovery rate.[13]

Rivers' new career was abruptly halted on Christmas Day 1892 when he was taken ill with scarlet fever and carted off to the London Fever Hospital. Returning two months later, he found a replacement doing his job, and was sent instead to help Dr George Savage give lectures on mental diseases at Guy's Hospital. Since retiring as physician superintendent of Bethlem four years earlier, Savage had built up an extensive private practice, thanks more to his genial social manner, robust common sense and passion for outdoor sports than to any penetrating scientific insights. (Years later, he would provide the model for the harsh and unsympathetic doctor in Virginia Woolf's *Mrs Dalloway*.)[14]

Modern historians have condemned the regime at Bethlem for the overuse of restraint, the insensitivity shown by male doctors to the needs of women patients, and the way that social class influenced treatment: thus a 1905 medical textbook declared that the powerful drug hyoscine should seldom be employed on 'persons of the higher social grades' but is useful with chronic patients, 'more particularly in patients belonging to the lower classes'. Rivers' objections (which would surface later) were slightly different. After the rarefied atmosphere at Queen Square, he found both the hospital's lower status and the lack of intellectual curiosity he encountered something of a comedown. British psychiatrists saw their task as being custodians of the mad and, although there was some contact between asylum doctors and neurologists, they tended to pursue different careers. With some exceptions, asylum doctors did not conduct academic research, as their

German and French counterparts did. And it seems that Rivers had by that time realised that his own talents lay more in research than in clinical medicine.[15]

But Rivers' concerns went beyond matters of status. He also questioned the nihilism – the assumption that nothing could be done to help the patients – then prevailing in British psychiatry. Reviewing *The Pathology of Mind*, by the leading British authority Dr Henry Maudsley, he questioned the author's underlying belief that it was best to 'disregard the psychological aspect of mental disease and to look solely at its material aspect as brain disorganisation'; such an attitude, Rivers argued, would only 'retard the study of insanity' in Britain. He also took issue with Maudsley's assertion that 'psychology had no terms for "these examples of decomposed will"':

> Dr Maudsley does not seem to recognise the possibility of a scientific psychology to be developed side by side with our knowledge of brain structure, which may be mutually helpful, each advance throwing light upon the other.

Rivers contrasted the 'loose classification and metaphorical description' which was the hallmark of British psychiatry, with the situation in Germany, where it had proved possible to impose coherence on this 'most confused branch of medical knowledge' by basing it on a 'true systematic psychology'.[16]

That was written later. In 1893, with nothing else on the horizon, Rivers signed on for a further six months at Bethlem. But then he was rescued: Professor Michael Foster, the impresario of Cambridge medicine, asked him to come and lecture on the physiology of the special senses to medical students. Although it was a tenuous position, Rivers immediately resigned from Bethlem. Before going to Cambridge, he went to Berlin to get an update on recent work in this field. There he worked briefly with the German psychiatrist, Emil Kraepelin, whom he admired as 'one of the few investigators who combine the study of experimental psychology with the care of the insane'; significantly, their research was on the effects of fatigue on mental health.[17]

It was hard going for the next few years. To earn a living Rivers had to teach in both Cambridge and London, which meant much commuting, and he was not a natural lecturer. 'He was rather nervous about it, and did not like it', a Cambridge colleague later recalled. 'This was partly owing to a hesitation of speech, which at times was quite embarrassing when he was speaking without notes. So he wrote out his lectures pretty fully.' Furthermore, the academic atmosphere was quite hostile. When the philosopher James Ward had first tried to set up a psychology lab in Cambridge, back in 1877, he was accused of 'trying to insult religion by putting the human soul in a pair of scales'. Much had changed in the intervening twenty years but there remained some hostility toward the new discipline. One elderly don referred to Rivers in the University Senate as a 'Ridiculous Superfluity'.[18]

Rivers' task was to distil for Cambridge medical students the findings of German experimental psychology, which over the previous two decades had developed into a scientific discipline using laboratory techniques to measure the processes of the mind and the senses. He focused especially on theories of vision and championed the work of the physiologist Ewald Hering (Henry Head's tutor in Prague) so vigorously that his students thought he had himself studied with Hering. He also carried out experiments on optical illusions, using equipment specially designed for him by Charles Darwin's son, Horace. He gradually came to be recognised as an authority in the field and was asked to contribute chapters to medical textbooks. His thoroughness and personal qualities, rather than any great brilliance or originality, also won him respect and recognition in the university. In 1897, he was granted an honorary Cambridge MA and appointed to the newly established lectureship in physiological and experimental psychology.[19]

Having finally won some recognition in Cambridge, it was understandable that Rivers should initially have turned down Haddon's invitation to go to the southern hemisphere. Why then did he change his mind? Rivers referred to the death of his mother, but there was another reason. His friend, the physiologist Lewis Shore, later recalled that Rivers 'was not completely happy or satisfied with his Cambridge life; and he had periods of depression and signs of nervous strain'. Poor health would have been an element in this, but Shore may also have been alluding to the mental conflicts brought about by Rivers' growing awareness of his own homosexuality, which he kept firmly repressed.[20]

Rivers had always used sea voyages to restore his health. This one was to change his life.

When Rivers had a change of heart and asked to join the Torres Straits expedition, Haddon agreed at once. But it left his team even more lopsided: he now had four doctors in his party, three of whom were psychologists. There was a further complication when McDougall, who did not wish to be subordinate to Rivers, offered to skip the Torres Straits leg of the trip and join the expedition in Borneo. Haddon would have none of it and insisted that McDougall come to the Torres Straits.

With time running out, Haddon now concentrated on getting to sea. It was a close-run thing; until the very last moment he did not know whether he would have enough money. The £300 already secured from Cambridge University and the £125 promised by the British Association would not cover the costs, but in December 1897 he managed to raise a further £370 from the Royal Society and £150 from the Royal Geographical Society. However, the Royal Irish Academy gave only £35 and the press magnate Alfred Harmsworth (who had recently donated £6,000 to Antarctic research) turned him down flat.[21]

Early in 1898, Haddon booked six passages on the steamship the *Duke of Westminster*. Seligman would sail by himself in January. Myers and McDougall, having passed their medical exams, now began to prepare themselves seriously. Myers struggled to assemble the equipment necessary for all the tests which Rivers and Haddon were planning on carrying out – cameras, recording machines, psychological apparatus – on a budget of just £25. In the end he paid for much of it out of his own pocket. Wilkin, the experienced traveller, was more laid-back, suggesting that he prepare some coloured prints of his English photographs 'so as to give the "niggers" an idea of what things look like over here'; he also thought 'an assortment of footer shirts' would 'trade like anything'.[22]

There were different responses, too, when Haddon required team members to sign an agreement, pledging all anthropological finds to Cambridge University and embargoing any account of the expedition before Haddon's own version had appeared. Rivers and Seligman were happy to sign but McDougall and Myers quibbled and insisted that Haddon limit the embargo to a year.[23]

Back in October 1897, before Rivers had come aboard, Haddon had written a short statement of the expedition's mission for the Royal Geographical Society, from which he hoped to raise money:

I have secured the following colleagues. Mr S. H. Ray, the greatest English expert on the languages of Oceania and especially of Melanesia and New Guinea . . . He will devote himself to the languages and phonology of the natives.

Drs W. McDougall and C. S. Myers – both qualified medical men and able physiologists – will study the medical and physiological sides of native life and more particularly will make experiments on the senses and sensibility of the natives. Practical psychology has never been studied in the field on savage and barbaric peoples – so this will be quite a new departure in anthropological research.

Mr A. Wilkin will pay attention to sociological facts – descent of property, land tenure, relationships and the like. He will also be mainly responsible for the photography.

It will be my business to see that no department of anthropological fieldwork is neglected and I shall lend a hand all round. Probably I shall chiefly study religion, art, handicrafts etc. and the psychical anthropology of the natives.[24]

Now the time had come to realise these intentions.

4

'We are all getting on famously'

RMS *Duke of Westminster* steamed past Ushant on 12 March 1898; Biscay and Gibraltar followed, in fine weather. A week after leaving London, the ship finally began to head east and the passengers could see the Sierra Nevada and the suburbs of Algiers.

The Cambridge team had unfinished work to do. 'All the party have been very busy reading shop [i.e. technical, as in 'talk shop'] books,' Haddon noted. Rivers corrected the proofs for a chapter on vision which he had written for a medical textbook and got Myers and McDougall to read and review it. Wilkin was set to work abstracting government reports on New Guinea. Haddon himself, always an early riser, began each day with a chapter of the textbook by the high priest of German experimental psychology, Wilhelm Wundt, before going on to correct the proofs for his own latest book. He also had time finally to do the expedition's accounts. He discovered that he was £150 adrift, but was confident he could make it up.[1]

There were a few diversions. On St Patrick's Day the sailors and the parsons gave a concert and dance for the steerage passengers and Haddon sat between some Irish girls. He was amused to see that the sailors sang sentimental songs and the parsons nautical ones. Some days later, as the ship passed Malta, a grander entertainment was given in the music room by the second-class passengers: Miss Mead played 'Alice, Where Art Thou?' on the piano, some of the gentlemen sang comic songs, and Myers contributed Ravel's *Bolero* and some popular airs on his violin; he thought the other performers 'lamentably wanting in quality'. Also on the bill was a short lecture by Haddon on the natives of Torres Straits, illustrated by lantern slides worked by Anthony Wilkin. It seemed to go down well.

On 22 March, the ship neared its first stop, Port Said. 'We are all getting on famously', Haddon wrote to his wife, Fanny. 'They all like Ray as he is so genuinely enthusiastic. I am very fond of Rivers and we get on capitally together. Wilkin and Ray are in the same cabin and they hit it off very well.' The following day the young male passengers helped the ship's crew to take on coal and then went ashore, still early in the day, led by Anthony Wilkin, who had been to Port Said before. 'After some light breakfast in a café,' Myers wrote in his journal,

> we engage an unusually persistent guide and bid him conduct us to 'where there are some fine women'. We pass up a side street from which we turn off into a narrow alley. Here our guide opens a narrow green door . . . various coarse prematurely aged women are roused from their slumbers to greet us. They troop into our room, looking more sleepy than we are. [There are] women of various nationalities; the pimp, a stout German woman, is anxious that we should pay to see the Can-Can danced, or at least to have some nearer relations to her charges. Disgusted with their tired and worn appearance, we walk out and follow our guide to a store where we buy pith helmets.

The young men had got there too late. The coming of electric light, which enabled ships to go through the Suez Canal by night, had had a disastrous effect on Port Said. 'Before that,' G. W. Steevens recorded in *Egypt in 1898*,

> every passing vessel had to lie the night at Port Said, and those were the days – or nights – of café concert, and roulette-table and dancing saloon, bands crashing from dusk to dawn, and gold flowing in torrents into the lap of Port Said. In those nights a second engineer could start out with sixpence in his pocket, and come back at sunrise with five hundred pounds in this pocket and a knife in his side. Now all that is gone. The shopkeepers and pimps have only a brief sunshine to get their hay in while the demons coal ship; and the faster the demons coal, the faster the money is aboard again and hull-down out of sight of Port Said.[2]

Myers and Wilkin settled instead for a donkey ride through the Arab quarter. Like many foreign visitors of the day, Myers was both excited and revolted by Egypt. 'At the present day there seems no pretence of morality. Everything is done to swindle the traveller and to obtain money from him in return for the most disgusting exhibitions', he complained. 'Again and again a child of ten will run up to me and says "Gentlemen, gentlemen, you will see photos, French photos, smutty, my good photos. You can see. Just pass the time." I went in to see a set of them. For sheer filth, devoid even of the most vulgar arts, they must I am sure be unrivalled.'

Rejoining the ship, the travellers progressed towards Aden, the weather now very hot and hammocks de rigueur. Myers relieved the monotony of the voyage by playing an unpleasant practical joke on Sidney Ray: 'I tell him that sandy white spots in his hand are strongly presumptive of measles. My opinion is to his mind independently confirmed by the ship's surgeon, Rivers and McDougall and he absents himself from church to avoid infecting the congregation. He even offers the surgeon his pulse.'

At the end of March they anchored off Aden. 'Several loads of boys approach us, many of whom clamber up into the deck, swearing like Christians and as keen for money as any Jew', Myers recorded. 'Two boat-loads of squalid-looking Jews came on deck, bargaining with feathers. In the steerage some Arab traders had set up a store, selling clothes and cigars.'

The arrival of a first batch of letters from home had left Haddon feeling guilty. 'Duty is a hard task-mistress, especially for women', he wrote to Fanny. 'What shall I say to you my beloved! What can I say that we have not said again and again and what I hope we shall say many a time in the future?' He did his best to reassure her, adding 'I am confident however that I am on the path of duty and I hope the work done will be recompense for the separation and labour.' All the party were well and 'we are getting on capitally together'. But Haddon did begin to voice some doubts about his team. Charles Myers reminded him somewhat of Fanny's brother, Trevor: 'there is the same lack of superabundant vitality. Life seems to have no joy for him. I am more afraid of his health than of anyone else's.' Sidney Ray 'has not yet developed much powers of conversation apart from linguistics; but at night he is very keen about the stars and that helps'. Against that, 'I am increasingly fond of Rivers.'[3]

Haddon still wasn't sleeping properly, but he had begun to relax.

Wundt's psychological lectures were now put aside in favour of Thomas Hardy's *The Mayor of Casterbridge* and other, more popular, novels and chatting with fellow passengers. On 6 April, nearly a month after setting out, with the passengers 'all now stewing in our juice', the *Duke of Westminster* reached Colombo and there was boarded by Singhalese washermen and tailors, from whom Haddon ordered one suit of khaki, two suits of ordinary ducks and one for dress, for all of which he paid thirty-four shillings. They were ready when he returned to the ship later in the day. A week later, passing through the straits of Sunda, the ship took on board a middle-aged Dutchman known as the King of Krakatoa who, as Myers reported, had 'bought the island a few days before the terrific eruption in 1883 which distributed the greater part of his property as dust over the four quarters of the globe'.

Then at last came a chance to escape the tedium of the ship. Learning that the *Duke of Westminster* would be held up in the Javanese port of Sourabaya, Haddon and his party rushed ashore at Batavia, and spent a night in hotels there. McDougall and Myers, sharing a room, enjoyed Javanese baths and had a long discussion about Dutch colonial policy over dinner with a British official from India. The following day they all took a train along the north coast of Java, giving Haddon a chance to take copious notes on the locals' use of scarecrows, techniques of rice production and methods of house construction, before passing another night in Sourabaya, which Haddon found 'the most interesting town I have yet seen in the East on account of the diversity of native types that one sees there'.

Excited by their shore foray, the Cambridge group rejoined the ship. While they were away, the crew had broached several cases of invalid port. 'General mutiny seems to have prevailed during the carousal', Myers noted. 'Things are now rapidly righting themselves. Penitence, more or less assumed, sits on the faces of the offending sailors.'

Only a week of the voyage now remained. Haddon's lecturing skills were still in demand. The Bishop of Queensland had already persuaded him to give a talk about 'the evolution of primitive religions' – during which Haddon 'was careful not to wound any susceptibilities'; now the passengers pressed him to perform again. He chose to talk about the 'bull-roarer', the bits of wood on strings used by Aborigines to generate a noise with which to communicate. 'None of those who lived in Australia had ever heard of it before!' Haddon wrote with disgust. 'My experience is that colonials know next to nothing of the natives of the country in which they live.'[4]

What did the expedition's members expect to find and do when they reached the Torres Straits? The prospectus for the expedition was very casual, by modern standards. Alfred Haddon had simply written: 'Experimental and observational psychology have received but scant attention in this country, and the psychology of the lower races has been totally ignored by us.' His better-informed sponsors would have understood that to mean that the expedition's purpose was to test out the proposition that primitive men, being lower on the evolutionary scale than civilised ones, had different sensory responses – that they had better hearing, eyesight and sense of smell but were less intelligent.[5]

The reference to 'lower races' is a reminder that Haddon shared the assumptions about race common to his era. Although the word 'race' was used in a very confused way by the general public and often meant nothing more than 'nation' or 'people' – thus in popular discourse it was common to hear talk of 'the English race' or 'the Irish race' – among scientists, the word 'race' referred to the peoples of the world and, at time when the British Empire was at its zenith, the existence of a hierarchy of races, with whites at the top, was taken for granted.[6]

When it came to race, Victorians believed in the natural inequality of human beings and were willing to generalise freely about the character of racial and ethnic groups. Modern historians have talked of 'scientific racism' during this period. In fact, race was science; in the biological sciences, the question of race defined the agenda – along with evolution. All the great men of late-Victorian science firmly believed in a hierarchy of races; none more strongly than Darwin, whatever his modern whitewashers would have us believe.

In 1870, his populariser Thomas Henry Huxley provided a classification scheme of racial types based on skin colour, hair colour and texture, eye colour, skull shape, and body stature. He identified five main races: Australoid; Negroid, 'Xanthochroi' (fair whites of Europe); 'Malancroi' (dark whites of Europe, North Africa, Asia Minor and Hindustan, including the Irish, Celts, Bretons, Spaniards, Arabs and

Brahmins); and Mongoloid (including the peoples of Asia, Polynesia
and the Americas).[7]

The islands of the Pacific had proved particularly contentious for
racial theorists. 'The issue at stake', the historian Douglas Lorimer
has explained, 'was the distribution, origin and affinities of those
groups identified as black (Australian Aborigines, Papuans, Melanesians
and Negritos) and those identified as brown (chiefly Polynesians, but
also Indonesians and Malaysians) . . . One side argued that the brown
and black peoples shared a common, probably black ancestor, and
subsequent change in physical characteristics had occurred most
commonly by migration and intermixture with other groups, and
rarely by geographical isolation and adaptation. The other side argued
for a classification of Papuan and Polynesians (or more broadly blacks
and browns) as two distinct groups sharing no physical and linguistic
affinities except in cases where obvious intermixture had occurred.'
Augustus Keane, the professor of Hindustani at University College
London, and a respected authority on race, claimed that the Polynesians
were descended from an earlier 'Caucasian' race in Malaysia.[8]

The purpose of anthropology, of studying 'lower races', was to
gain insights into the way human evolution had worked. And the
development of medical anatomy had provided a powerful new tool.
'The interest attaching to the study and examination of the brains of
the lower races of mankind, is briefly summed up in the phrase, "brain
as an organ of mind"', the anatomist Humphrey Rolleston declared
in 1888. 'What material differences are there between the brain of an
educated moral man and that of a sensual animal-like savage? What
correlation is there between the physical conformation of the cerebral
hemispheres and the mental development of their owner?' According
to Rolleston, the average weight of the male European brain was 49
ounces, the average weight of the Negro brain about 44.3 ounces, and
that of the 'primitive Australian' considerably less. Anatomy, however,
could only take the enquiry so far. Other methods must be used to
understand the primitive mind – such as 'psychology'.[9]

In fact, the idea of applying psychology to primitive peoples had
been around for a while. Back in 1876 the Victorian sage Herbert
Spencer had drawn up a programme for mapping out 'the comparative psychology of man' which addressed three subjects: the 'degree
of mental evolution of different human types', the relative mental

natures of the sexes in each race, and the more special mental traits distinguishing different types of men. Spencer was vague as to the methods to be employed, but two decades later German experimental psychology and Francis Galton's work on mental testing in England had provided a whole battery of mental tests and the equipment to carry them out with.[10]

The Germans had made psychology scientific – and enabled the new discipline to gain authority and distinguish itself from its academic neighbours, philosophy and physiology – by finding ways of measuring vision, hearing and touch and, then, mental activity. The great German experimentalist Hermann von Helmholtz believed, for example, that the measurement of human reaction times enabled him to make deductions about the nature of otherwise unobservable neural processes. The British, as William Rivers well knew, were very much passengers in this scientific race, following in the Teutonic wake. The one important thing they brought to the discipline was their empire: with a network of missionaries, traders and administrators all over the globe, they could more easily translate the idea of applying experimental psychology to 'primitive people' into practice.

According to the historian Graham Richards, Haddon was 'absolutely fascinated by race and the relations between peoples both in the present and in prehistory'. He also had his own theories. In *The Study of Man,* the book he corrected in proof on his way to the Torres Straits, he spoke of three primary races of mankind arising by evolutionary divergence from some unknown ancestral stock, each primary race becoming 'arrested' or 'specialised' in different directions. Haddon had created his anthropometric laboratory in Dublin to 'help differentiate racial types in the Irish population and so assist in the study of the settlement of Ireland' and the purpose of his ethnographic survey of Britain had been 'to establish the racial history of the United Kingdom using evidence from physical anthropology, folklore, dialect studies, history and archaeology'. One of Haddon's reasons for returning to the straits was to study the languages used there more scientifically. But he was now equally interested in psychology – by which, as James Urry has written, he meant 'the scientific investigation of the physiology of sense perception to elucidate the mental processes of man. He believed that these processes might vary in different races due to differences in their stage of evolution.'[11]

Although William Rivers had not worked in anthropology, he was the nephew of James Hunt who, besides being a pioneering speech therapist, was a strong believer in racial hierarchies. Hunt did not even believe in monogenesis and told a scientific meeting in 1866 that 'There is as good reason for classifying the Negro as a distinct species from the European as there is for making the ass a distinct species from the zebra . . . if we take intelligence into consideration, there is a far greater difference between the Negro and the European than between the gorilla and the chimpanzee.' After Hunt's premature death, Rivers had been offered his uncle's anthropological library but had declined it because he was not at that time interested in the subject.[12]

The other members of the expedition also had their own views on race. William McDougall would later describe himself as 'half Celt and half Saxon', Charles Seligman had done medical research on the inheritance of eye colour and hair type, Charles Myers had published on physical anthropology, and Anthony Wilkin in his letters to Haddon expressed robust late-Victorian views, referring to the islanders as 'niggers' and 'blackamores'.

This, after all, was the era of Rider Haggard.

PART TWO

ON THE ISLANDS

5

Murray Island

At dawn on 23 April 1898, the *Duke of Westminster* dropped anchor at Thursday Island, the administrative centre of the Torres Straits archipelago. Charles Seligman came on board to welcome his fellow explorers to Australasia. Since getting to Melbourne in early February, he had done some natural history in New South Wales and a little anthropology in Queensland.[1]

Haddon found Thursday Island 'still the same assemblage of corrugated iron and wooden buildings which garishly broil under a tropical sun, unrelieved by that vegetation which renders beautiful so many tropical towns'. To Seligman the town was 'like a collection of magnified sardine boxes turned on end'. Although the pearl fisheries had by now been depleted, the island was more prosperous than a decade before, with stores, lodging houses and thriving two- or three-storey hotels. Its population of some 3,000 souls consisted largely of foreigners, the bulk of them now Japanese, 'much to the disgust', Haddon noted, 'of most of the Europeans and Colonials'. Their main complaint was that 'the Japanese beat the white men at their own game' by working harder and living more cheaply.[2]

Haddon's party were enthusiastically received by the local officials, many of whom remembered him from a decade before. They also learned that the Queensland government had made a very welcome donation of £100, which nearly closed the expedition's funding gap. It took a week to unload all their equipment and to organise transport on to Murray Island, at the eastern end of the straits, which Haddon had chosen as their first base. 'I shall be glad to get the party there as here we are eating our heads off – doing nothing', Haddon complained. To pass the time, Myers played tennis and engaged a Malay named Charlie Ontong as the expedition's cook. He proved to be a fair cook and claimed to be 'all straight now since his woman died' but then went missing and returned a day later saying 'Me go booze', having spent half of the £4 of advanced wages he had been given.[3]

Finally, on 29 April, the expedition members began loading cargo on to the *Freya*, a ketch belonging to a Dane called Anderson, but there was so much gear that a second ship, the schooner *Governor Cairns*, also had to be hired. The following day, Haddon, Rivers, Ray and Seligman set off in the *Freya*, having had to rout Captain Anderson out of Burke's Bar, leaving Myers, McDougall and Wilkin to follow.

The wet season was unusually protracted that year. It rained hard and the seas were rough. Both parties made difficult passages. 'There was some little bother in getting across to Murray, our first headquarters – finally we were nearly all wrecked', Wilkin wrote home. The *Freya* dragged her anchor at night among the reefs and on the *Governor Cairns* a flying jib broke and the vessel ran aground on a coral reef off Stephen's Island. 'The skipper was plainly frightened', Myers noted. 'We lost two kedges and an anchor and were busy hauling on tackles from 11 p.m. to daylight.' The following day the schooner was refloated and the voyage resumed.[4]

Another hazard was that the young scientists were unused to the tropical heat. 'All of us suffered a good deal from heat burn on the feet having imprudently exposed them to the sun and salt water', Haddon recalled. Seligman 'lay on deck like a piece of wet blotting paper and let the seas break over him without stirring'. Rivers, ever the psychologist, noticed that the pain of sunburn vanished while he was 'in imminent danger of shipwreck. So long as the danger was present I moved about freely, quite oblivious of the state of my legs, and wholly free from pain.'[5]

When Haddon's party finally reached Murray Island on 6 May they were grateful for a hot bath and a meal at the house of John Bruce, the schoolmaster and magistrate on the island. Haddon immediately began renewing old friendships with Ari, the 'mamoose' or chief of Mer, and Pasi, the chief of the neighbouring island of Dauar. 'We walked up and down the sand beach talking of old times, concerning which I found Pasi's memory was far better than mine', Haddon reported. He found an abandoned Mission residence which, though slightly dilapidated, would make an admirable base for the expedition and began unpacking equipment there. Sidney Ray and William Rivers, still suffering from sore feet, had to be carried up the hill.

The following Sunday, Haddon greeted the congregation after the service and invited the islanders to a reception to show them photographs from his earlier visit.

> We had an immense time. The yells of delight, the laughter, clicking, flicking of the teeth, beaming faces and other expressions of joy as they beheld photographs of themselves or of friends would suddenly change to tears and wailing when they saw the portrait of someone since deceased. It was a steamy and smelly performance, but it was very jolly to be again among my old friends, and equally gratifying to find them ready to take up our friendship where we had left it.

The next morning an immense heap of bananas and coconuts arrived at the mission – a welcoming present from the islanders.[6]

By the time McDougall, Myers and Wilkin turned up two days later, the expedition's work was well under way, though it took a further week to settle down and unpack. Haddon's plan was to spend a short time on Murray Island to get the psychological testing programme started and then hand it over to Rivers while he went off to New Guinea. Haddon used two strategies to get the islanders involved in the project. First he took pictures of those who had been measured. When news of this got around, people poured in to have their 'pikki' taken. At the same time, 'the first thing we did after arranging the house was to convert a little room into a dispensary, and very soon numbers of natives came to get medicine and advice'. By 10 May, Haddon reported, the surgery was 'in full swing – many patients including some children come to be treated and the Drs are very happy. Rivers [whose feet had swelled up] is lying in the next room to the surgery and as he is lying in bed he tests the patients for colour vision. He is getting some interesting results.'[7]

A week later, Haddon was satisfied with the progress made. 'We all pull together harmoniously and we all appear to be pretty well acclimatised', he wrote to Fanny. 'We are now in trim for constant work.' The main problem, however, was the irregularity of the visits of the natives: 'sometimes they come in batches, at other times not at all'. But 'they are very friendly and I'm glad to say my party get on well with them'. Haddon finally went off to New Guinea on 23 May, taking Sidney Ray, Charles Seligman and Anthony Wilkin with him. His intention was to clarify the relationship between the Torres Straits islanders and the Papuan people of New Guinea and to carry forward his growing interest in the decorative art of New Guinea. Ray's language skills would be an important tool.

Rivers, Myers and McDougall were left to get on with the testing.[8]

Murray Island – or, more correctly, Mer – was about five miles in circumference, hilly, made up of lava and ash and thus very fertile, producing coconuts, yams, bananas, and sweet potatoes to support its 450 inhabitants. The islanders supplemented this diet with fish and an occasional turtle. The only white man there was John Bruce, the teacher and magistrate, 'a quick, refined, observant Scotchman of the working class', his chief rival being Finau, the Samoan minister of the London Missionary Society. 'This Samoan is a troublesome fellow', Myers wrote. 'He comes among the people telling them how Samoans are all the same as white men and had Christianity before the white man visited them.'[9]

It took a little while to adjust the flow of subjects for testing but eventually Rivers and his colleagues arranged with Bruce and the local chieftain to have a limited number of men come every morning. On the whole, the islanders were pretty cooperative – rather more so than their English equivalents would have been, Rivers thought. The rationale of the project, they were told, was that 'some people had said that the black man could see and hear, etc. better than the white man and that we had come to find out how clever they were, and that their performances would all be described in a big black book so that everyone could read about them'. This incentive seemed to work, though it meant that there was less enthusiasm for tasks that were long and repetitive or in which the islanders felt they were failing.[10]

The setting brought other difficulties. For one thing, years of sea diving had so damaged the eardrums of most of the adult men that there was little point in comparing their hearing to that of 'civilised' people. More importantly, much of the team's equipment did not adapt well to local conditions. It wasn't just that Myers' violin rapidly fell apart in the humid atmosphere; much of the psychological apparatus, for all the 'great searchings of heart as to what apparatus to take out and which to leave behind', did not work. Some instruments had been damaged in transit; others had been designed for sterile laboratory conditions and were not robust enough for tropical temperatures. Lovibond's Tintometer, for testing colour perception, did not live up to its maker's claims; Politzer's Hörmesser, for assessing keenness of hearing, was useless in the open air against 'the rustle of palm leaves and the breaking of the surf'; and Zwaardemaker's Olfactometer, for measuring smell sensitivity, was no good because the islanders 'entertained a great objection to inserting the glass tube within their nostrils'.[11]

The plan was that Rivers would test vision; Myers, hearing and taste; and McDougall, touch and skin sensitivity. Members of the team went about their work in different ways. McDougall tended to devise simple experiments and then quickly to adjust them to his subjects' levels of tolerance. For instance, when he found that pricking the islanders' skin in many places was straining their patience, he 'confined his observations to the skin of the forearm and the nape of the neck'. In testing the pain threshold he used an algometer (an instrument which presses a point against the skin with various levels of pressure):

> I instructed my subjects to cry 'Stop' at the moment they began to feel any pain, but I found that in nearly all cases it was possible to detect the onset of pain by observing the slight flinching which it causes in the expectant subject. The onset is a perfectly sharp and definite change in the sensation.

One of McDougall's most interesting experiments was designed to establish whether the islanders' blood pressure while doing physical and mental work was similar to that of Europeans:

> For, since the effective working of the brain is so intimately dependent on a rapid circulation of the blood through it, and since the circulation is so largely determined by the state of the arterial blood pressure throughout the body, the power of mental activity to raise the general blood pressure must be of great importance in promoting the vigour and effectiveness of mental processes. And it may be that this power

is an element of fundamental importance in determining the superiority of the higher races. It has been frequently asserted that the
inferiority of the black races is due to the cessation of the growth of
the brain at an earlier age than in the white races, and it may be that
that this is in part, or wholly, due to a less active response of the blood
pressure to mental activity.

To get his subjects' minds going, he gave them a pencil and a maze
drawn on a card, which he claimed quickly got them involved. He
then measured their blood pressure with a sphygmomanometer, just
as a nurse would today. He found only minor differences in the rise
of blood pressure during mental and muscular work between Murray
Islanders and Englishmen. So that any difference in intellectual capacity
was not caused by this.[12]

Myers had the most difficult task. As very few of the gizmos he
had brought for testing hearing and smell worked on Murray Island
and, of those that did, several were far too delicate and complicated,
he and Rivers had to spend much time trying to devise substitutes.
Testing reaction times was then very much at the cutting edge of
experimental psychology, and the Torres Straits team probably only
bothered to do it because a recent paper on 'reaction time with reference to race' had caused a stir. 'That the Negro is, in the truest sense,
a race inferior to that of the white can be proved by many facts,' the
American psychologist R. Meade Bache had written in 1895, 'and
among these is the quickness of his automatic movements as compared
with those of the white . . . the Negro is in brief, more an automaton
than the white man is.' The laboratory technology for measuring both
visual and auditory reaction times was much too sophisticated for use
in the field and Myers had been obliged to devise simpler alternatives,
which were less sensitive. In fact, however, the islanders liked the
visual reaction test because they were told it was their speed in hunting
that was being measured: 'they seemed clearly to understand that I
was somehow able to find out whether they were likely to be good
or bad marksmen by studying the rapidity of their reactions', Myers
wrote. They would ask how their results compared with those of
friends or rivals. The results appeared to show that 'the average auditory reaction of the young Murray Island adult and the young English
townsman are almost identical', but 'his visual reaction appeared to

be distinctly longer'. But, as Myers acknowledged, there were many possible explanations for this.[13]

In testing vision, Rivers had the easiest task. Eyesight tests of the sort familiar today had already been developed and the apparatus for testing colour blindness and differences in brightness – colour wheels and different coloured wools, designed by Ewald Hering and other experimenters – was comparatively portable and easy to use in the field. Rivers found that, although the Murray Islanders could make out birds or a canoe on the sea from great distances, this was not because their sight was better but because their eyes had been trained to pick out details. And, when their perception of colour was tested, they proved capable of exceptionally fine discrimination between different samples of coloured wools, yet their vocabulary for naming colours was crude and limited.[14]

Gradually the scientists began to be interested in the natives, not simply as experimental subjects but as human beings. As the weeks passed, they learned pidgin and began to communicate with the locals. The Murray Islanders, Charles Myers wrote, were 'an intelligent, unreserved, excitable people of medium stature, with dark frizzy hair, a chocolate-coloured skin and Papuan features, which the most favourably impressed European could not term handsome'. His first impression of them, in church on the first Sunday, had not been flattering:

> The average Sunday attendance is probably about fifty. On work days the men wear calico cloths around their hips; many wear a thin vest. On Sundays the kilt is replaced by trousers. Coats are worn, helmets by three or four and there is a profusion of brilliantly coloured scarves and handkerchiefs. The women put on equally gaudy clothing. Boots are rare, but when a wearer of them enters church, he makes for the farthest seat, taking care to let his envied possession be seen as he walks with no light step along the aisle. Most of the women carry a baby and towel. I took the latter to be a kind of family pocket handkerchief, forgetting that the native babies, like European babies, would have their faults that they had not yet learnt to control.
>
> The singing is very harsh and of a terrible volume, the time is bad and the effect like a German brass band. They love hymn singing: most of them join in. A few younger lads and lasses giggle, converse, whether

by mouth or eyes, or even play cat's cradle. But the behaviour is on the whole remarkably good. On this, the first day of my attendance at church, a curious sermon is given – curious because of its polylinguistic character. It is preached by a Raratongan missionary . . . in Motu Motu and translated, sentence after sentence, by another visitor into Samoan, the teacher finally giving it to the Murray people in Meriam [Murray Island dialect].

Myers noticed that the island was changing. Many of the younger men had gone to Thursday Island to dive for pearl shell and bêche-de-mer while older people clung to the simple ways in which 'all that is necessary to support life is a little gardening'. When Myers asked one man about meals, he replied 'sun he come up, sun he go down, eat and drink all day before missionary come. Missionary he make him eat, breakfast sun there, dinner sun up here, and supper sun down there. We go sleep midnight. We get up along sun. We go sleep sometimes two, sometime three hours, sun up here. Suppose we tired, we sleep longer.'[15]

'The character of the natives appears to be as diverse as it would be in any English town', Myers concluded, when he had got to know some of them. One of their cooks, Jimmy Rice, was 'the sturdy, plodding thick-set workman, with probably a touch of the British Tar about him'; whereas Debe, the other, was 'a genius. His originality is even amusing.' Myers considered Pasi, the mamoose of Dauar, to be 'the best example of the high-strung nervous type: his excitement when he is telling a story is remarkable'. But the most interesting character was Ulai, an old man who hung around the scientists' camp and helped them in various ways. Ulai was a rogue, 'the personification of cunning', who would utter untruths with 'twinkling half-closed eyes, uplifted brow, lobbing tongue and almost fevered smile'. When Ulai discovered that Rivers was interested in collecting anthropological artefacts, he began to produce various treasures of his own and eventually, with a great show of reluctance, sold him the stone on which his sexual prowess was based; only then did he also produce a collection of rods recording his sexual conquests. Now that he was an old man he no longer needed it, he said.[16]

Among the fifty-seven boxes of luggage the expedition had brought to Murray Island was an Edison phonograph, for recording music on cylinders. As Myers became more sympathetic to the natives, he took to his task of recording their music with greater enthusiasm. The children had been taught European songs in school, and sang them with remarkably correct intonation, but the adults were still more used to what Myers called 'Papuan airs'. He also realised that several of the islanders he had tested composed their own songs, the drum being the only instrument by which they were ever accompanied. On 16 July he had 'a great morning of music' when 'Joe Brown', an elderly man who was reputed to be the best singer on the island, finally agreed to be recorded. Joe proved a poor singer to Myers' ear, but his example prompted several other men to join in, and Myers was able to fill seven cylinders with dance and semi-religious songs. When, however, he asked the men to sing the song which had accompanied Bomai, the initiation ceremony Haddon had witnessed a decade earlier, they declined, 'excusing themselves on the ground that the two boats which have left here today will capsize and will be given over to the sharks if the Bomai song is uttered'.

It was now almost two months since Haddon and his party had gone off to New Guinea and those left behind on Murray Island were beginning to show signs of strain. The diet was monotonous: the excellent Malayan cook Charlie Ontong had left with Haddon, and all efforts to get their Murray Islander servants Debe and Jimmy to make bread had failed (higher wages having no effect). Occasionally there was turtle on the menu, which Myers considered 'indifferent steak' and, attending a native feast, he was offered 'a fowl, a pig's head, some soup and a variety of vegetables. "Soup" is a word given to any meat well boiled with a fair quantity of water. We have it nearly always daily. The most inviting dish is one of yams cut up and cooked with grated cocoanut and cocoanut-milk.' By now it was also raining incessantly and the explorers were plagued by insects: 'We have had an increasing influx of ants, flies, grasshoppers, mosquitoes, flying beetles and pests as the season has advanced or, perhaps, as the brutes have spread the knowledge of our arrival to each other', Myers wrote. He was also 'horror-stricken, at the close of a visit to the W.C. to find that a number of small maggots [ate] the excreta. Naturally, I have ever since the discovery been chaffed by Rivers and McDougall with tender enquiries after the "worms". It turned out that the wretched things had crawled up from the bottom of the pan.'

The monotony was lifted when Haddon and his party finally reappeared on 20 July. 'They have had none of the rain which visited us,' Myers wrote resentfully, 'and they return sneering at our poor fare of tinned meats and condensed milk after feeding on the fat of the land (fresh milk, cream, fresh meat) with the Roman Catholic fathers at Yule.'[17] Haddon had been moving around in New Guinea at his usual headlong pace, cadging lifts on government schooners and hospitality off the missionaries, recording information about tribal customs and practices, buying curios and artefacts, sketching the tattoos on women, observing house construction methods and initiation rites in passing, pausing longer whenever dances, sorcery, or children's games were involved. The party had been on a three-day trip on horseback into the Astrolabe Mountains, waded through crocodile-infested estuaries and eaten wallaby-tail soup. Haddon had collected samples of native hair, stone clubs, knives made of boar's tusks and a sorcerer's kit; he had played ludo with an archbishop; and attended Catholic Mass for the first time in years.[18]

Trailing along in his wake, Sidney Ray, Charles Seligman, and Anthony Wilkin did their best to carry out observations, pick up language, and take photographs. Wilkin was bewildered by it all. 'I have just come back from a four days' trip into the bush with a couple of horses and a native "boy"', he wrote to his Cambridge tutor, Oscar Browning. 'I must confess to have got more experience than pleasure out of it, but I saw treehouses and lived with the natives so that their point of view became conceivable. We are a little too apt to look for embryo Anglo-Saxons among blackamores.' He tried to provide a summary of native social and political organisation in New Guinea but found it hard: 'the more one knows about it the more involved it becomes', he wrote, adding, 'I can't help thinking we expect too much definiteness from savage minds.' Wilkin kept his doubts to himself, but eventually Seligman put his foot down and insisted on staying in one place for a time. As he was paying his own expenses, Haddon could not prevent him but his annoyance was profound.[19]

Soon after his arrival back on Murray Island, Haddon reviewed the expedition and assessed his colleagues in a letter to Fanny:

Rivers is extremely pleased with M.I. & the natives – he thinks it an ideal place for psychol study & he has done a lot of work that will

prove of great interest and value. I think you will be very pleased with
his report. Myers has turned out well – slow but very careful and
accurate & he is getting quite keen & will turn out good work.
McDougall has not risen to the occasion. Wilkin is not very satisfactory,
he is too young – Ray works well but he is very limited. Seligman is
very keen but I don't like him and will try to shunt him.[20]

How had McDougall 'not risen to the occasion'? Haddon himself had
been away for two months, so the phrase – much used by disappointed
vicars and disobeyed housemasters – must have come from Rivers.
The fragments of evidence make it clear that McDougall behaved
arrogantly, did not do his bit in the general tasks, and devoted the
minimum of intellectual effort to his share of the testing. He himself
wrote later that he had 'greatly enjoyed the time' on Murray Island
but had 'accomplished very little' there. At this time, on his own
admission, his 'youthful arrogance continued unabated', which made
him a tiresome companion. He had always regarded the Borneo leg
of the expedition as the more interesting and had wanted to avoid
the Torres Straits altogether; the rather cursory reports he later wrote
on the islanders' cutaneous sensations, muscular sensations, and variations in blood pressure suggest that his approach was indeed designed
to minimise the intellectual labour involved. Myers' diary records that
McDougall did not find the islanders, with their frizzy hair and 'Papuan'
(i.e. Negroid) features, very sympathetic. He may also have adopted
a rather colonial attitude to the expedition's servants.[21]

But maybe the work itself was not to McDougall's taste – repetitive, boring and often frustrating as it was. His hero William James
had written that German experimental psychology 'taxes patience to
the utmost, and hardly could have arisen in a country whose natives
could be *bored*'. The great Harvard psychologist thought there was
'little left of the grand style about these new prism, pendulum and
chronograph philosophers. They mean business, not chivalry.'[22] What
we know of McDougall's later attitudes and behaviour suggests that
he quickly lost patience with the work and came to question the whole
premise of the expedition – applying modern experimental psychology
to primitive peoples. It appears that once again he had joined a team
and then failed to show team spirit; and that mattered more on a
tropical island than in a college rowing eight at Cambridge.

There is no record of what was said between the two men, but soon after Haddon's return, McDougall went off to Darnley Island to attend to the ailing Captain Harmon, a Conradian figure who, ruined by the failure of three successive banks, had cut himself adrift from his family and come out to the straits. He was now living alone on the island. McDougall therefore missed the high point of the whole expedition.[23]

Ever since his first visit to the islands, Haddon had been obsessed by Malu, the secret ceremony initiating the young men of Murray Island into manhood. It had been his ambition to witness and record the ceremony – and partly for this reason he had brought with him a bulky 35 mm film camera which, after many mishaps and misadventures, had finally reached the expedition. Now, somehow, he had to persuade the old men on the island briefly to throw off their Christian modesty and revert to the old customs; make them defy the church and, in particular, their Samoan preacher. It became a trial of wills. When Haddon and Myers, with all their bulky equipment, went to the open area where the ceremony had once been staged they found no one willing to participate, and spent a gloomy night in an overcrowded hut nearby, surrounded by old men who were frightened of losing their positions in the church and prayed loudly through the night.[24]

But Haddon, Myers noted, was 'bent on getting what he wants'. He produced his secret weapon, an old Malu drum, of enormous length measuring four feet eight inches, mobilised the white magistrate and 'spoke strongly to some of the influential men'. When Rivers, Ray and Wilkin arrived, they too began telling every native in the area that the old rites were going to be revived one more time. Haddon's pressure worked. As Myers noted:

Finally when the Malu drum begins to sound, the impulse to leave all work for Malu is irresistible. Even the most pious islanders succumb to the appeal; one by one they turn up on the scene. We wait until a signal is given that all is ready; we then repair to a spot in the bush cleared of the bamboo trees which surround it. Here the old ceremony of initiation was held. Here the youth to be initiated into manhood, Kesi as they were called, first saw the dread masks of Malu and Bumai. Today the ceremony is performed which no white man ever witnessed before.[25]

Haddon's description of the ceremony itself is imprecise: 'Grotesque masks worn by ruddled men, girt with leafy kilts, as they emerged from the jungle, and very weird was the dance in the mottled shade of the tropical foliage, a fantasy in red and green, lit up by spots of sunshine.' Afterwards the visitors went down to the beach and witnessed a series of dances being executed by the numerous clans of the island.

As the sun set, Haddon's party returned home after a very tiring but most exciting day. McDougall, returning from Darnley Island, woke them up at midnight.[26]

6

'Oppositions of opinion'

When Alfred Haddon returned to Murray Island from New Guinea on 20 July 1898, he had to make a quick decision. While he had been away, the party on Murray Island had received a letter from Charles Hose warning that the team needed to reach Borneo not later than September in order to avoid the rainy season. This had caused Myers much anguish. 'The expedition is essentially a Torres Straits expedition', he wrote in his diary. 'Borneo was an afterthought. But its finances clearly will not allow of our stopping on here if we give up Borneo. To give up Borneo is to break a promise.' Now, with the mailboat due to leave the next morning, a decision had to be made. Haddon held a council of war and, after considerable discussion, ruled that Myers and McDougall would head off to Borneo with all the apparatus in a few weeks, leaving Haddon, Ray and Seligman to follow two months later when their essential work was finished. Rivers and Wilkin would return to Cambridge then.[1]

It took two weeks to organise a boat but finally one was found. As the time to leave approached, Myers became positively sentimental about the place and the islanders. They, too, seemed sorry to see him go. 'Poor old mamoose wipes away seemingly genuine tears when he realises my departure. We have been great friends. He has learnt all my names, calling me by any one of them that suits him at the time. He is a delightfully simple and honest fellow.' After several days of loading all the gear aboard a cutter, they were ready. 'Farewell to Murray Island', Myers wrote on 24 August 1898. 'Farewell to its inhabitants, so little touched by the artificial hand of European civilisation.' After two days at sea, living off potted meats and bananas, Myers and McDougall reached the comforts of Thursday Island and, over a cold supper at the Grand Hotel, began planning their trip.[2]

When Alfred Haddon had accepted Charles Hose's invitation to call in on Sarawak (or British Borneo), no serious thought had been given to the logistics involved. No one seems to have realised that, although the Torres Straits and Sarawak might seem to be quite close on the world map, in travelling time they were almost as far apart as Britain and Australia. Now, on Thursday Island, Myers and McDougall had a nasty surprise: they learned that there was no shipping service between the nearest big port, Batavia (a Dutch colony), and Singapore (a British one); so that they would first have to travel north to Hong Kong and then back south to Singapore before catching a boat to Sarawak – a detour adding 3,200 miles to the journey. It took them four weeks and two ships to reach Singapore (via Darwin, Timor and Hong Kong), and they then had to spend a further week in the Raffles Hotel waiting for the small cargo boat, carrying about a dozen saloon passengers and over 200 Chinese coolies for the plantations, which took them on the final leg of the voyage, from Singapore to Sarawak.

Six weeks' travelling together will test any relationship. It destroyed the friendship between Charles Myers and William McDougall. Somewhere on the long journey, tempers frayed and angry words were spoken. McDougall had clearly been warned to behave himself; writing to Haddon from Thursday Island, he had expressed the hope 'that we shall avoid "oppositions of opinion"'. But it was Myers who cracked, finally driven to the end of endurance by McDougall's arrogance and self-centredness. When or how it happened is not clear. Myers' journal provides no clues.

Finally, the travellers reached Kuching, the capital of Sarawak. On the morning of 4 October 1898 they anchored off the Baram River and in the afternoon crossed the bar and tied up at the mouth of the river itself. Myers was startled when a Dayak tribesman in a loincloth appeared in his cabin asking for a light and a drink. The sound of the frogs and mosquitoes reminded him of bassoons and piccolos. The following day they weighed anchor at dawn and, as the sun rose, passed up the long mud-brown sluggish river, three to four miles wide, its banks lined with huge trees. Only a few brilliantly coloured kingfishers, parrots and snow-white cranes could be seen. After ten hours they reached Claudetown, guarded by an impressive newly built fort. There Charles Hose came to meet them. 'We stroll over to Hose's house which he has built himself', Myers recorded. 'He intends going into the far interior soon. We are to go with him.'[3]

Charles Hose was the second son of a Norfolk clergyman. While he was still an undergraduate at Cambridge, his uncle, the Bishop of Singapore, had procured him a job in the administration of Sarawak and he had gone to the East, aged only twenty-one. Sarawak was one of the stranger outposts of the British Empire, created in the 1840s when James Brooke, a British army officer turned adventurer, had wrenched a substantial piece of territory from the Sultan of Brunei and turned it into a family-run British colony. His nephew and successor as 'rajah', Charles Brooke, was 'both an English gentleman and an oriental despot' ruling with 'benevolent authoritarian paternalism'. The younger Brooke outlawed the slave trade and ended slavery by 1886, and made progress in reducing the traditional practice of headhunting among the Dayak tribesmen. As a colonial ruler he was cautious, encouraging some economic development among the Malay and Chinese population on the coast while allowing the native tribes – Dayaks, Kayans, Kenyahs and Klementans – to continue their traditional ways of life more or less untouched; he was proud that Sarawak had been spared the exploitative excesses of King Leopold's Congo. The 'white rajah' ruled through a small, hand-picked group of English administrators and insisted that his officers be 'gentlemen' and not marry until after ten years' service, though he did not object if they became involved with local women.[4]

Charles Hose rapidly established himself as an effective administrator with a passionate interest in the local people, animals and plants. Myers noted that his exceedingly comfortable house contained a very large room for repairing and storing biological specimens – as well as quarters for his Malay wife and three half-caste children. For over a decade Hose had been in sole charge of a substantial and remote area which he ruled as a personal fiefdom. He had invited the Cambridge team to his district, partly to share with them his genuine passion for anthropology, partly to have an audience for a diplomatic coup he was hoping to pull off.[5]

Hose's method of governing had always involved incessant travel among the tribes he administered, to hear their concerns and impress them with his own personality; but he had long been planning a more ambitious expedition, deep into the interior, near to the border with Dutch Borneo, to make contact with a tribe known as the Madangs, who had been raiding and terrorising their neighbours. Hose wanted to negotiate a general peace and to bring the Madangs within British rule: to persuade them, if possible, to attend a general peace-meeting at Claudetown, at which the outstanding feuds between them and the Baram folk might be ceremonially washed out in the blood of pigs. To achieve this it was vital that he took with him members of other tribes hostile to the Madangs.

On 9 October 1898 Hose and his party set off, travelling in some style on the Resident's launch with a retinue of six Sea Dayak rangers and two policemen, and towing half a dozen boats, including one for their use upriver. After visiting several villages on tributaries of the Baram, they continued up the main river until it became necessary to abandon the steam-launch and take to the boats. The river was swollen by recent rains and the travellers could make only slow progress by repeatedly crossing the stream to seek the slack-water side of each reach. Several times they had to camp out in the open.[6]

For Myers and McDougall, it was exhilarating to spend days travelling up a huge muddy river bordered with dark impenetrable jungle under a tropical sun, pausing periodically to visit tribesmen; or it was terrifying and uncomfortable. But matters were complicated by the fact that Hose, who belonged to a particular English tribe himself – clergyman's son, minor public-school boy, and passionate Freemason – took an immediate shine to McDougall, the Highland Celt, and an evident dislike to Myers, the urban Jew. And, whereas McDougall had been somewhat detached on Murray Island, he was now fascinated and engaged by the 'very interesting and likeable tribes of the interior of Borneo', with their attractive, fine-featured, intelligent-looking faces, and magnificent physiques. 'I greatly enjoyed wandering in wild places among primitive peoples and found it easy to make sympathetic contacts with such people', he later wrote.[7]

Myers, by contrast, was ill at ease. Doubtful about having come on to Borneo in the first place, he was a nervous traveller. He continued conscientiously to record the sights, sounds and social customs – it was not true that 'Bornean youths were compelled to go headhunting before girls would accept them in marriage'; 'Ignorant travellers' had got cannibalism wrong: 'What they do is cut chunks off the dead and offer them as sacrifices to the hawk, their bird of omen which has given them hope in the expedition.' But a note of vexation crept into his journal. He disliked the local people, whose reserve and self-control unnerved him after the open friendliness of the Torres Straits islanders, and he hated the noise produced by 'thirty miserable-looking half-starved dogs per house'. The 'stamping and shouting' accompanying toast-drinking with locals and the activities of the pig population caused him further unease. On 15 October 1898, Myers felt compelled to 'pen a few lines to describe the troubles of defecation':

> The innocent traveller wanders forth after breakfast in search of a suitable spot for the operation. But however far from the house he may go, he is certain to be followed by three or four pigs. These animals grunt and sniff at a most disrespectfully short distance while he is endeavouring to perform Nature's duties. A long stick is often necessary to keep off the boldest of them. Indeed a dorsal eye would not at time be amiss to prevent the intrusion of a snout into the man's arms. When the operation is over, the pigs rush up and greedily swallow the faeces. How much this distracts from the pleasures of defecation and subsequent pig-eating.[8]

Myers' discomforts eased a little when the party arrived at the substantial dwelling of Tama Bulan, the most famous and powerful chief in Sarawak, whose assistance would be crucial to Hose's mission. The old man, whose 'expression of shrewdness and capability' reminded Myers of Lord Rosebery, the recent British prime minister, urged his followers to accompany Hose on his journey and to assist him in his mission of peace; but before they could leave, days had to be passed in feasting, making speeches and toasts, and taking omens on the journey. Favourable omens having been observed, 'sacrifices of pigs and fowls were offered on the altar-post of the war-God and the various rites needful to complete the preparation for a long journey were performed'. At last, on 23 October Hose's party, by now numbering over a hundred, 'resumed the toilful ascent of the main river'.[9]

'The water is very bad today', Myers wrote on 26 October. 'We have a terrible struggle past the ends of tree-trunks which project up along the riverside. Rain comes on. Water frequently dashes into the boats. The men work manfully, literally lifting the boat over the stones with their long poles. At each rapid we rise from three to nine feet further above sea level. We reach the village of Long Liam in the afternoon. It is a filthy place surrounded by a pig-sty.' The following day, they were some 230 miles up the river, having travelled 170 miles from the fort and reached an altitude of about 700 feet above sea level. 'We enter the river Akar, a tributary of the Baram', Myers wrote. 'The rapids are very bad. Water frequently dashes over into the boat. At one o'clock we reach the house where we are to spend the night.'

That night, Myers developed fever. He was taken to the hut of the local head man, and lay there for five days, his temperature reaching 105°. It was decided that he would have to go back to Claudetown, and that McDougall must accompany him. Myers was full of guilt: 'I have spoiled McD's pleasure', he wrote in his journal.[10]

It took Charles Myers almost a month fully to recover his health. For a week in December he convalesced and then spent some time 'attempting to get people to be investigated psychologically', but he found the locals in Sarawak suspicious of the whole exercise and reluctant to be tested, whatever stratagems he employed. On the other hand, it was comparatively easy to record their music, which Myers found much more sophisticated than the primitive drumming of the Torres Straits islanders, with 'Mongolian, Malayan and possibly Indian and other influences'. Their songs, often sung at dances associated with headhunting, were accompanied by an instrument called the *keluri*, a mouth organ made up of six bamboo pipes attached to a gourd wind chamber. Myers described the sound it emitted as a 'drone'.[11]

His thoughts were now turning to home, and to the houseman's job waiting for him at St Bartholomew's Hospital in London. Seeing little point in staying longer, he boarded the first home-bound steamer to come along, even though it meant leaving the South Seas without saying goodbye to Haddon. A letter would have to do. Writing to him on 16 January 1899, Myers tried to make light of the problems in Borneo: 'my stay has been a delightful one'; Hose had been 'extraordinarily kind and attentive'. He also asked for permission to publish a popular article on Borneo on his return to England. But he could not hide his feelings towards McDougall:

I have done my best to maintain a friendly understanding between McD and myself. I regret that once my irritability got beyond my control, provoked by his cultured egoism . . . I have never met a man endowed so much with strength of brain and body & so superficially with the qualities of good fellowship. My disappointment in one, to whom I long looked up as the ideal of what I myself could be, is very great.[12]

Eventually Haddon replied. 'It is unfortunate that you and McDougall did not hit it off better,' he wrote, 'but enforced companionship for a long period usually results in a certain amount of friction and misunderstanding which wear off as the occasion passes unless foolishly or perversely perpetuated.' In a letter to Fanny he added, 'so far as I can make out, Myers was very petty and Hose did not think anything of his work. Thought he was too casual . . . I am very glad M[yers] has left. H[ose] did not like him and it is perhaps just as well that S[eligman] & M[yers] have not met here.'[13]

What, then, had Haddon and his party been up to in the meantime?

'The fountain of knowledge'

Back in the Torres Straits, William Rivers had been enjoying himself. With Myers and McDougall and their cumbersome psychological equipment gone, he could finally concentrate on his own method of interacting with the locals. In his Cambridge research on vision, Rivers had looked at family histories, to see whether abnormalities in eyesight were hereditary. He now discovered that, if he asked Murray Islanders whether particular characteristics such as keen eyesight ran in families, he could find out all sorts of other information about their lives. Moreover, the natives proved to be surprisingly well informed about their family connections, often going back several generations. Rivers soon became fascinated by this method and by the end of his stay had tabulated the genealogies of every native of Murray Island going back as far as could be remembered.[1]

But this information had to be carefully extracted. 'A tremendous amount of secrecy had to be exercised in these enquiries in Murray Island,' Haddon recalled, 'and one never knew in what corner or retired spot one might not come upon the mysterious whispering of Rivers and his confidant. The questions one overheard ran mostly in this wise, "He married?" "What name wife belong him?" "Where he stop?" "What piccaninny he got?" "He boy, he girl?" "He come first?" and so forth.'[2]

Complexities had to be negotiated. An islander might have two names or casually assume a new one or have been married several times, often to widows with children. The most confusing issue of all was the widespread custom of adoption on Murray Island, so that in many cases children did not find out who their parents were till they reached adulthood; often they never knew. Rivers was also well aware that here the system of naming relationships was very different from that back in England. For example, the sisters of someone's mother, that is, his maternal aunts, were also called 'mothers'. Rivers found that he could operate with only five terms: those for husband, wife, child, sibling, and 'proper' father. He would get from one man the names of his parents, siblings, and their families, and then repeat the process with his father, and so on. Much varied sociological information could be obtained in this way. Rivers had not simply discovered an important new tool: he had broadened the expedition's focus, beyond the basic interest in ceremony and religion, into sociology.[3]

In the meantime Haddon had continued his mixed programme of research and Seligman had travelled on his own, in New Guinea and around the islands. In September, Haddon, Rivers, Ray and Wilkin moved on to Kiwai Island, off the coast of New Guinea, to stay with Haddon's old friend, the missionary and so-called 'Livingstone of New Guinea', James Chalmers. Haddon was irritated not to find Seligman there, as arranged. 'I do not know when we shall meet,' he wrote to Fanny, 'and I have virtually decided not to take him on to Borneo with Ray and myself. He has been of no use to the Expedition proper and no one likes him.' But a few days later Seligman turned up and Haddon was forced to admit that he had efficiently gathered a wealth of anthropological material. 'I have had a clearing-the-air talk with him and think all will be right for the future', he wrote to Fanny. 'I can't say I look forward to having him as a travelling companion in Borneo, but as I expect he will turn his attention to anthropology seriously and as he is very keen and an ardent collector I feel it my duty to stick to him – so as to give him additional training and experience.'[4]

This new buoyant mood was apparent when the party moved on to their next destination, Mabuiag in the western Torres Straits, where Haddon had spent two months on his first expedition. Haddon had always liked the place, finding the atmosphere there more relaxed than on Murray Island. The weather was cooler and the white men found they could get about with more comfort. The work was 'going on swimmingly and without friction', Haddon reported.

Rivers continued to develop his genealogical method on Mabuiag. In one respect it was harder because the families there were larger and until recently polygamy had been widespread; but whereas he had been obliged to operate with great secrecy on Murray Island, on

Mabuiag the social climate was quite different: the information was obtained in public and doubtful points in genealogy frankly discussed by native men and women. This enabled Rivers to work much more rapidly and to cross-check information more easily. Haddon applied Rivers' method himself on the small island of Saibai and was able to complete a genealogical census of the population before midday. This, as the historian Ian Langham has written, 'seemed to indicate that the technique provided an incredibly accurate means of exposing the fundamental speciation and workings of preliterate society'.[5]

Excited by Rivers' work, with Seligman now 'well in hand', and 'even Wilkin interested in getting information', Haddon began to feel more relaxed. 'I have at times been very despondent in the past', he admitted to Fanny on 25 September. 'Now I feel the load is off my shoulders – there is not much more to do.' But Haddon had other worries – about money, getting to Borneo, keeping up morale back in Cambridge, and whether to apply for jobs in England. Even with the Queensland government's £100, funds were now running low and, as Haddon reckoned he needed another £200 for the trip to Borneo and the return voyage to England, Fanny was kept busy getting the expedition's treasurer, Professor Macalister, to chase up the British Association's £125 (which had never materialised) and applying for a second grant from Cambridge University. If neither of those were forthcoming, he told her, she must borrow from her brother Harry, adding, unconvincingly, 'I hope you have no financial worries.' The uncertainty continued until the end of the year when Haddon finally heard that Cambridge University had voted him another £150.[6]

Haddon tried to sustain his wife's morale with frequent, loving and supportive letters. Mostly, his faith sustained her. 'I just live from one mail to another', she told him. 'As soon as one letter is received and gloated over my whole being is in expectation of the next.' But even Fanny, the perfect wife, had her darker moments. One of her letters forced Haddon to deny that they were drifting apart and to promise that in future 'we can live a more joint life by doing things together'. As the moment to leave the Torres Straits and make the journey to Borneo approached, Haddon's guilt and doubts intensified. 'Often I wish that I wasn't going to Borneo,' he confessed to Fanny, 'but when that trip is over I know I shall feel that I ought not to have missed it for anything.'[7]

In late October Rivers and Wilkin caught the boat back to England, as planned; and on 15 November 1898, Alfred Haddon, Sidney Ray and Charles Seligman began the journey to Borneo. While at sea Haddon wrote a long and guilt-ridden letter to Fanny, to reach her by her birthday:

> For the last few days I have especially wanted you. I suppose that the lack of outside interest that characterises a sea-voyage has revealed the aching want in my breast. When I am working hard my time is too occupied to brood over my cravings – though I always feel a need of you and instinctively think of things in relation to you – wonder how you would like the situations or the people and wish that you could share my pleasures and new experiences – though not the trials and discomforts. It will be good, my beloved, to resume a joint life once more – it is not good for me to be alone and we have had more than our fair share of separation. We have each tried to do our duty – yours in patient self-renunciation and in bearing single-handed the cares of the family. I know your brave and loyal selflessness will bear its sweet fruit not only in your experience, but as an example to our children and to other wives. I too have tried imperfectly though it has been to do my duty – but I have had the charge of excitement and the exhilarating quaffs of freshly drawn draughts from the fountain of knowledge – and it is stimulating to drink where none have drunk before. But I, too, have had my periods of depression – those sluggish backwaters of life, when joyous activity ceases and all is covered with an unwholesome scum. How often have I yearned for your serene common sense and your sweet comfort! Now we can hopefully look forward to our reunion which draws nearer and nearer. I can only send you thoughts and wishes to arrive on your birthday. What present can I give when you have my life – my all!

It had taken Myers and McDougall six weeks to get from Thursday Island to Baram; it took Haddon nine. His party had arrived at Kuching on 12 December only to find that the bar at the mouth of the Baram River was now impassable. They had to linger in the town over Christmas and New Year and then take a ship to Limbang, to the north of Borneo, and from there proceed by a long overland detour to Hose's base at Claudetown. The travellers made the best of the situation: Ray got on with his language studies, Seligman saw friends, Haddon visited museums and called on the Sheik of Brunei in Limbang. But this further delay was agonising for him, especially when he heard that his mother had been dangerously ill for months and his relatives had kept the news from him. In his letters to Fanny, his rationalisations for the expedition became more tortured. He now argued that he had done Fanny a favour by leaving her alone in Cambridge for so long, forcing her to reveal her true courage and strength. Quite how she responded to being told that 'The world will be mentally the richer for our separation' we can only imagine.[8]

From a professional point of view, there was a danger that when Haddon finally reached Baram he would not have enough time to do any worthwhile fieldwork. He told himself that Charles Hose would be coming to England on leave the next year and could be pumped for information then; and, to relieve his sense of failure, sent numerous artefacts to Cambridge. The vice chancellor and professors Ridgeway and Macalister each received 'a hairy Kenyah shield, blowpipe and spear' – useful weapons to wield in the University Senate – yet the cost of passage of these objects had to be met by his sponsors and would prove an important factor in damaging his relationship with Macalister, the expedition's treasurer.

Finally, on 23 January 1899, from the head of the Malinau River, Haddon was able to report that: 'We have here got to the last stage of our wandering.' Charles Hose had had to go off into the interior, taking McDougall with him, but his deputy met Haddon's party and escorted them across country and down the River Baram, to Claudetown. 'I can scarcely express how relieved I am to find myself here safe and sound after nine weeks' meandering', Haddon wrote to Fanny.[9]

When Hose returned from his journey, Haddon was happy to let him take the lead. 'We had a warm welcome from Hose and now everything goes on greased wheels. Hose has taken a great fancy to Ray, I am glad to say and S[eligman] is put into a back seat which will do him a lot of good.' The Resident – 'a tremendous man both as to size and energy – as keen as possible and generous to a fault' – had decided that the Cambridge group should accompany him to Mount Dulit, in the south of his domain, to observe the natives and the flora. 'S stops behind to do folk medicine etc – this will be a great mental and moral relief to me as he has got on my "nerves". Hose can treat him splendidly & has arranged one or two little excursions for him by himself', Haddon reported to Fanny. For all Hose's enthusiasm, Haddon was getting very homesick. 'Our trip has more or less knocked us all up', he wrote to his wife. 'I am very tired of knocking about and long to be at home and at rest again.' When it came to making the final climb to the summit of Mount Dulit, Haddon 'did not feel equal to the job' and left McDougall to complete the ascent while he sat in the tropical sun writing.[10]

Before leaving, Haddon and McDougall were able to witness an extraordinary spectacle. Charles Hose's journey up the River Baram the previous October, accompanied some of the way by McDougall and Myers, had been designed to persuade the tribal groups in his region to come together for a formal peacemaking. After almost a month of travelling he had finally arrived at the Madang district and, after a nervous period when the omens and auguries were tested, brokered a peace between the Madangs of the neighbourhood and those from Baram that would be confirmed by a ceremony at Claudetown the following year.

Now, in April 1899, long war canoes full of warriors began to arrive at Hose's base; and eventually there were 6,000 of them staying in leaf and palm shelters on the open ground in front of the fort. There was some concern that no member of one group, the Tinjar, had appeared. Had they taken offence for some reason, or met with bad omens? Were they taking the opportunity to sack and burn their neighbours' undefended houses? But finally they too turned up, in full war dress, 'with feather coats of leopard skin and plumed caps plaited of tough rattan'. Immediately a noisy battle began, a sham fight in which the grievances of one side were symbolically acted out. But Hose was careful to keep the newcomers well apart from the rest.[11]

The following day 6,000 tribesmen began to assemble in a great conference hall of palm-leaf mats which Hose had constructed for the occasion. When everyone had gathered, the sight of all their old enemies was too much for some of the natives and they flew at them. The great Tama Bulan himself was wounded in the ensuing struggle, and stood with blood pouring from his face; it seemed as if real hostilities were about to begin. But Hose, plucky as ever, intervened to point out that the two guns of the fort were trained on the conference hall and Tama Bulan, always the statesman, urged restraint on his own people. The crisis passed.

Attention now moved to the boat race scheduled to take place the next day. There were over twenty boats, each with some sixty or seventy men sitting two abreast, and it took some time to get them into line, but when the starting gun went off, 'the twenty boats leaped through the water, almost lost to sight in a cloud of spray as every one of those twelve hundred men struck the water for all he was worth'. McDougall, an oarsman himself, reckoned the 'rate' was about ninety a minute, forty being a good rating on the Thames Tideway. Very soon, two boats drew away from the rest. 'It was a grand neck-and-neck race all through between the two leading boats, and all of them rowed it out to the end', McDougall recorded. The winners were 'a crew of the peaceful downriver folk, who had learned the art of boat-making from the Malays of the coast; and they owed their victory to their superior skill in fashioning their boat, rather than to superior strength'.[12]

Hose and his party, following the race from their launch, now began to worry. 'How would the losers take their beating? Would the winners play the fool, openly exulting and swaggering?' Fortunately, however, the winners 'behaved with modesty and discretion' and Hose diverted attention by taking the launch through the main mass of canoes – forcing them to cope with the launch's wash.

Having broken down the hostility Hose made sure that everyone returned to the conference hall. After more reading of the omens and extracting of pigs' livers, the Resident himself was called upon to give his opinion. Hose, McDougall reported, was

> soon able to show that the only true and rational reading of the livers was guarantee of peace and prosperity to all the tribes of the district; and the people, accepting his learned interpretation, rejoiced with one accord.
>
> Then the Resident [Hose] made a telling speech, in which he dwelt upon the advantages of peace and trade, and how it is good that a man should sleep without fear that his house be burnt or his people slain; and he ended by seizing the nearest chief by the hair of his head, as is their own fashion, to show how, if a man break the peace, he shall lose his head.

Feasting followed and – the next morning, while the tribesmen still had hangovers – the payment of two dollars per head in taxes. Haddon wondered what might come of this gathering – 'a Sarawak nation may in time arise, composed, as practically every European nation is, of several races and innumerable tribes'.

On 20 April 1899, after a stay of almost three months, Haddon and his party left Claudetown. Five days later they sailed from Kuching on the long voyage home. Finally, on 31 May, after fourteen months abroad, they landed in England.

Haddon had successfully brought off his expedition. But what had it achieved?

'A lasting memorial'?

When William Rivers returned to Cambridge from the Torres Straits in January 1899 he told everyone that 'the whole thing was a splendid success'. At the end of May, Alfred Haddon got back from Sarawak to a hero's welcome – the Royal Society elected him a fellow and put on display a collection of polished stone ornaments he had brought back. In September, the British Association for the Advancement of Science devoted an entire day of its annual meeting to the expedition and Haddon, Rivers, McDougall, Myers, Ray and Seligman all spoke, putting on a show of unity which delighted Charles Hose, who had by now also arrived in England. 'I am rather pleased that the little unpleasantness between Myers and McDougall has gone no further', he wrote to Haddon. 'It is best for all of us that we should keep our opinions to ourselves, unless in self-defence we are bound to speak. I myself shall say nothing to anyone as it is no affair of mine – I only wonder that we all got on so well together.'[1]

Cambridge University, however, was distinctly unimpressed by the expedition and it took Haddon some years of struggle to earn credit for his achievement. For all Fanny's efforts to hold the fort in her husband's absence, Haddon's position had in many ways declined. While he was away, his original patron, Professor Macalister, had brought in a substitute, W. L. H. Duckworth, a highly competent anatomist who proved more to his liking than Haddon. Macalister then convinced the university to create a post tailor-made for Duckworth, and Haddon was left with no alternative but to return to his position at Dublin. This further separation from wife and family was unwelcome and Haddon also found himself in financial straits, forced to borrow money from relatives to pay expedition bills. 'Being in Dublin [is] a great hindrance to me; but I have to earn in Dublin that bread that enables me to work at anthropology in Cambridge for nothing', he complained to J. G. Frazer. 'My prospects were never so unpromising for the last twenty years as they are at the present time.'[2]

Haddon's Cambridge supporters rallied to his cause. Frazer and William Ridgeway, the professor of archaeology, wrote a memorial to the General Board of Studies arguing that the university had a moral obligation to repay Haddon for the work he had done, largely unpaid, to establish anthropology in the university, 'at personal sacrifices which only those who know Professor Haddon's circumstances can appreciate'. This appeal secured widespread support among the professoriat, but the university remained unconvinced. Finally, in May 1900 it did establish a lectureship in ethnology with a meagre annual salary of £50. This gave Haddon a real dilemma. He could not live on the Cambridge salary; yet, increasingly, he felt the strain of shuttling back and forth between East Anglia and Ireland. His health had begun to suffer and, unusually, he gave way to despair: 'I certainly have need for all my philosophy,' he wrote to a friend, 'as I have had many disappointments in life and now have so much to discourage me. Unless I get something more in Cambridge, I must give up the struggle . . . give up Anthropology . . . and return to Dublin.'[3]

Then, in February 1901, he was given a life-line. Christ's College, Cambridge elected him to a fellowship. It was only for three years and the salary of £200 was not vast, but Haddon decided to accept. 'You might as well starve as an anthropologist as a zoologist', was Fanny's advice. Her faith was rewarded four years later when Haddon was given a senior fellowship at Christ's. He duly became university lecturer and reader in anthropology, deputy curator of the Archaeology and Anthropology Museum, and a major Cambridge figure, giving garden parties and speaking in the Senate. He was at long last able to reward Fanny for her years of support and to secure his children's future. But he never became a professor at Cambridge, and that may have had something to do with the writing up of the Torres Straits expedition.[4]

The uncertainty about Haddon's future made it doubly important that the expedition's findings be produced as quickly as possible. Haddon himself led the way: despite being forced to commute between Dublin and Cambridge, in 1901 he published *Head-Hunters: Black, White and*

Brown, a charming, humane and witty popular account of the expedition, closely based on his journal, to generally favourable reviews. But much more important were the reports of the expedition's scientific work, and in particular, of the psychological testing.

When the expedition members presented their findings at the British Association meeting six months after their return to England, Haddon made no summarising statement at all, just a presentation with slides. The task of giving an overall verdict on the tests fell to Rivers, who cautiously offered two 'general results'. Firstly, the experiments showed 'very considerable variability': 'It was obvious that in general character and temperament, the natives varied greatly from one another, and very considerable individual differences also came out in our experimental observations.' Secondly, 'the natives did not appear to be specially susceptible to suggestion but exhibited very considerable independence of opinion'. Myers, McDougall, Seligman and Ray then gave preliminary reports on their specific areas.[5]

However, Professor Macalister then pointed out that in order to assess the Torres Straits findings it would be necessary to have comparative material. And so the team set out to replicate their studies on a British control group. Rivers chose fifteen students of psychology and twelve schoolchildren from the village of Girton, Myers travelled to a small village near Aberdeen, and McDougall worked with the inmates of a convalescent home at Cheadle in Lancashire. In addition, Rivers tested the colour perception of eighteen Labrador Eskimos who were visiting England and went out to Egypt to test the colour vision of Egyptian peasants employed on archaeological digs there.

Despite the delay caused by extra testing, Rivers managed to produce the first volume, *Vision*, in the autumn of 1901 and Myers' and McDougall's accounts of *Hearing. Smell. Taste. Cutaneous Sensations. Muscular Sense. Variations of Blood Pressure and Reaction-Times* followed two years later.[6] The reception was generally favourable. Haddon's friend Francis Galton hailed the first volumes as 'a very important contribution to both physiology and psychology . . . a lasting memorial both of the activity of Cambridge anthropology and of the genuine character of the scientific spirit', but another reviewer wondered whether 'the islands of the Torres Straits were wisely chosen, since there are still to be found so many [other] races untouched by European habits, language, clothing, or religion'.[7]

What, then, did the expedition conclude? How did the senses of savages compare with those of civilised men? The methodological problems the scientists had experienced – with equipment, especially – meant that 'most of the research resisted confident interpretation'; indeed the whole premise on which the expedition rested – that the methods of German experimental psychology could profitably be used to study primitive people – remained in doubt. The historian Graham Richards has argued that a collective decision was taken to spin the material. 'The team members were, in truth, in rather a fix', Richards has written. 'They could hardly tell their sponsors that the whole exercise had been a debacle from which little could be positively concluded, nor would nascent British psychology's cause have benefited.' As a result, the reports 'became a virtuoso exercise in the art of writing up unsatisfactory research as positively as possible short of outright dissembling'.[8]

Richards overstates his argument. In fact, the reports frankly discuss the problems encountered and are tentative in drawing firm conclusions. On vision, Rivers found that 'the visual acuity of savage and half-civilised people, though superior to that of the normal European, is not so in any marked degree'. Their remarkable eyesight, recorded by so many travellers, was, he concluded, due more to the eye being trained to interpret what it saw rather than to any superior power of vision. He then drew a broader conclusion. The very closeness with which savages observed nature was, he argued, 'a distinct hindrance to higher mental development'. 'Over development of the sensory side of life' might, he thought, also account for the fact that 'the uncivilised man does not take the same aesthetic interest in nature which is found among civilised people'.[9]

When it came to colour vision, Rivers was concerned not only with how the islanders saw colours but with the words they used to describe them, finding that they combined a very sharp eye for colour differences with a very limited vocabulary for describing them. Adapting an argument made four decades earlier by William Gladstone, the British prime minister and classical scholar, he went on to argue that the languages of the four peoples he had observed in Oceania showed different stages in the evolution of colour terminology, which corresponded in a striking manner to the course of evolution which scholars had deduced from ancient writings.[10]

Myers' problems with equipment meant that his extended discussion of hearing, smell and taste drew hardly any firm conclusions. He did, however, find that the simple reaction times of the young Murray Island adult and the young English townsman were almost identical, and in more complicated tests there were differences which he attributed to 'racial differences in temperament'. McDougall was more prepared to generalise. He found, for example, that the sense of touch of the Murray men was 'twice as delicate as that of . . . Englishmen', while their susceptibility to pain was 'hardly half as great'.

Not only are the findings cautious, the writers also make few attempts to place the Murray Islanders within any broad scheme of human evolution. Indeed, there still remains some confusion among historians as to what the expedition's overall findings were. Were Murray Islanders different from civilised white men or essentially the same – in the modern phrase, 'just like us'? Gradually, over the next decade, a conclusion of sorts began to emerge among the expedition members – that there was no conclusion. But that, in itself, was significant. If the whole exercise had yielded no firm results, if this team, with all their expertise, had not found evidence of significant differences, then savages and civilised men must be roughly the same. That, in time, became the conclusion.

In fact, the expedition's greatest legacy was the impression it left on those who took part in it. 'I was tempted to make field-anthropology my main line: for I had greatly enjoyed wandering in wild places among primitive peoples and I had found it easy to make sympathetic contacts with such people', McDougall wrote in 1930. 'Looking back I cannot now understand why I rejected this alluring prospect. I remember that my conscious ground of rejection was characteristically arrogant. I said to myself, "That field is too easy for me"; and turned back to my original scheme of direct attack on the secrets of human nature.'[11]

In Myers' case, that temptation to change never arose; he never mentioned the idea. But his friend Charles Seligman did make the move and proved a great anthropologist: 'Sligs', much beloved and respected by all; not least by Haddon, their differences in New Guinea long since forgotten. It was on William Rivers, however, that the Torres Straits left the strongest mark: Haddon once said that the expedition's most important achievement was to turn Rivers into an anthropologist. On

24 April 1900, only a year after his return to England, Rivers appeared at the Royal Anthropological Society to promote 'the genealogical method' as an important new tool which 'enables one to study abstract problems, on which the savage's ideas are vague, by means of concrete facts, of which he is a master'. Yet the direction Rivers took in his new field would be decisively shaped, not by Haddon or Seligman, but by an Australian anatomist he had first met at St John's in 1896.[12]

In January 1901, Rivers was in Egypt. Having already tested the vision and colour sense of Labrador Eskimos, he now wanted to get a third set of results by testing the Egyptian labourers on an archae-ological dig which Anthony Wilkin and a friend were conducting in Upper Egypt. Since returning from the Torres Straits, Wilkin had taken a second-class degree in history, and, after failing to get a job as a war correspondent in the Boer War, joined the archaeologist Flinders Petrie in Egypt. Having learned the techniques, he was conducting a further dig with his friend, David Randall-MacIver.[13]

When Charles Myers also travelled to Egypt to recuperate from illness, an impromptu Torres Straits reunion took place. 'In the quad-rangle sat Rivers in his short-sleeves and flannels, just as in the old Murray Island times,' Myers wrote in a letter to Haddon, 'with a native intent on colourful wools before him. I spent a delightful five hours with them all.'[14]

This would be the last time that Rivers, Wilkin and Myers got together. Four months later Wilkin contracted dysentery while returning home through Cairo. He died suddenly, aged twenty-five. William Rivers also visited Cairo, to call on an old Cambridge friend of his, the anatomist Grafton Elliot Smith, now a professor in the Egyptian capital.

Their reunion opened a new chapter in Rivers' life.

PART THREE

MUMMIES AND
MELANESIANS

Hunting Platypus

One morning in August 1895, three young men set out to hunt duck-billed platypus in the Blue Mountains of Australia, the great range of peaks that separates Sydney from the interior of New South Wales. Pitching their tent on the banks of the Duckmaloi River, they sat up round the camp fire and rose with the freezing dawn to stalk their timid prey.

The platypus is not easy to hunt. It is a shy, retiring creature that seldom leaves the riverbank. Nor is it good to eat. Yet almost from the moment that the British got to Australia, they hunted the animal down: at first out of curiosity, because it seemed to be such a freak of nature, at once bird, reptile and mammal; then, after the publication of Darwin's *On the Origin of Species*, because of clues it might give to the nature of evolution.

In Australia, scientists first became aware of marsupials, mammals which do not have a placenta to the womb and carry their young in pouches; and of monotremes, animals such as the platypus, which perform the functions of excretion and reproduction through a single passage. As the skills of dissection improved, so the animal's secrets were uncovered. By the 1830s, it was known that the platypus had mammary glands which were seen to secrete milk; a decade later its internal reptilian characteristics were described. It was long thought that it was oviparous – that it laid eggs – but this was not finally confirmed until 1884 when the Scottish scientist William Hay Caldwell, conducting large-scale 'kill and purge' operations with droves of dogs and Aborigines, eventually shot a platypus which had laid its first egg and had a second in the mouth of its partially dilated uterus. He at once sent a triumphant telegram to the British Association for the Advancement of Science meeting in Toronto:

'Monotremes oviparous, ovum meroblastic', which his scientific confreres understood to mean that 'although the platypus had reptilian features and laid eggs from which its young were hatched, the eggs had large yokes that did not divide, like those of birds, and suckled its young like other mammals'.[1]

The three hunters gently probing the bushes on the banks of the Duckmaloi River eleven years later were all medical researchers. One of them was interested in the platypus's reproductive system and knew that in the month of August, in the depths of the Australian winter, the female would be carrying her eggs inside her, before laying them in the early spring. A second was investigating its muscles and nervous system. The third was after its brain; although only twenty-four years old, he already had an international reputation as an expert on the brains of mammals. His name was Grafton Smith.[2]

Grafton Smith was the grandson of a baker from the London district of Soho who had emigrated to Australia with his family in 1857. Grafton's father, Stephen Sheldrick Smith, had in his youth attended a school in Leicester Square and worked in Winsor & Newton, the artists' materials shop just off the Tottenham Court Road, before enrolling as a pupil at the Working Men's College which a group of Christian Socialists established in Red Lion Square in 1854; afterwards he claimed to have been taught there by the great John Ruskin. Rudimentary though this training was, it served Stephen Smith well in Australia. He became a teacher, rose quickly, and was the headmaster of a school in the small town of Grafton in northern New South Wales by the time his second son was born in 1871 – the boy was named after the township. A decade later, Stephen moved to another school in the Sydney suburb of Darlington. He gave his children some of his own sense of ambition and purpose. His son would later recall that 'his knowledge was wide and wonderful and encouraged me to cultivate a universal curiosity'.[3]

Grafton Smith's boyhood was shaped by the glories of the natural world observable on his doorstep and the theories offered by Darwin and his successors to explain them. When he was ten, he was fascinated by a physiology textbook and began to dissect sharks on the beach with his penknife. Walking to school in Sydney one day, he saw an evening course on physiology advertised and was soon being thrilled by Huxley's *Lessons in Elementary Physiology*. As a teenager he bought David Ferrier's *Functions of the Brain* and on a schoolboy visit to the university medical school he was allowed to handle human brains. The professor remarked that no one knew all the convolutions of the brain. 'I remember silently framing the vow that I should be the one exception to this statement', Grafton Smith later recalled.[4]

After an uneven passage through high school, Smith entered the medical school in Sydney. It was soon clear that his main interest was in doing medical research on the evolution of the brain – and on the brain in evolution. His superiors steered him away from cats and hedgehogs, which every laboratory in Europe was busily dissecting, and towards the distinctive Australian marsupials and monotremes. By the time he was twenty-four – and hunting platypus in the Blue Mountains – he had published articles in the leading European medical journals.

In 1896 he came to England and was quickly snapped up by the Cambridge anatomist, Professor Alexander Macalister. By the autumn of that year, he was a research student at St John's College, spending most of his time in the physiology lab, where he met and soon became friends with William Rivers. Here he continued his research. 'I expect to be at the brain work for a long time to come', he wrote. At first he stuck to the 'Marsupial and Monotreme stuff', comparing the nervous systems of southern marsupial moles, echidnas (anteaters), and bandicoots with those of bats, using specimens from the zoological collection in Cambridge or donated by the network of colleagues he was building up. A steady stream of papers and conference appearances brought him into the academic mainstream.[5]

The 'very jolly' life of a research student had to be modified in 1898 when his Australian scholarship ran out. Forced to earn a living, Smith 'endured a life of abject slavery' for three months, demonstrating in anatomy and physiology every morning, lecturing on the brain and coaching students in the afternoon and evening. 'You can imagine the agony of such a life to my lazy temperament', he complained. But he kept up the flow of papers – in September 1898 he and a friend were staying in Cambridge digs and 'working away like Trojans . . . We are expecting to hear of the Battle of Omdurman tomorrow.' The following year, Smith turned all his previous work into an 'essay', on the strength of which he was elected a fellow of St John's in November 1899 at the age of twenty-eight.[6]

By then, his work on the evolution of the brain had taken an important step forward. Through contacts with Huxley's old associates, he was introduced to the Royal College of Surgeons and was eventually invited to catalogue the college's immense collection of reptiles' and mammals' brains. Having mastered the neuroscience of exotic Australian animals, Smith now had the opportunity to investigate the nervous system of creatures higher up the evolutionary tree. 'I have examined the brain in almost every genus in the whole Mammalian and a large number of sub-Mammalian genera', he wrote in April 1900. He could now establish in detail how the simple 'smell-brain' of the more primitive creatures had evolved into the much more complex organ found in mammals, 'informed and modulated by later information coming in from all the other senses'; how instinct had become intelligence. The catalogue he eventually produced became a classic in comparative anatomy.[7]

When Smith first came to London, he struck a fellow dinner guest as 'earnest in his work, quiet, scientifically one-sided; anatomy evidently his one interest. Will be a big anatomist but not a big man. So reticent, giving no local colour of his life.' He obviously broadened out at Cambridge. Here a contemporary remembered him as 'shy and taciturn at first with strangers', with a drooping moustache which 'gave him an appearance almost of melancholy', but 'this soon vanished when he talked with friends on matters in which he was interested, when he became lively and humorous and the best of company'. Most of his friends were colonials, and he was aware that 'the research student from a university in our Dominions had not had the opportunity to obtain such a complete training as the Cambridge man, and there were obvious lacunae to be filled up'. The grandson of a Soho baker had his social insecurities, too; for which he compensated by developing a pontifical manner and using his full monicker; in England, plain 'Grafton Smith' flowered into 'Grafton Elliot Smith'.[8]

In July 1900 Elliot Smith was offered the professorship of anatomy in the medical school in Cairo. That September he left Cambridge, paused in London to marry Kathleen Macredie, a Sydney girl he had met in Ireland, and reached Egypt a month later.

Elliot Smith was charmed by Cairo. It was 'the gayest and most cosmopolitan city on the face of the earth', he wrote. 'It is also the most beautiful place and intensely fascinating.' He quickly got the university's anatomy department up and running and planned to spend three months of the year in Europe. He told friends that he had quite resisted the temptation to dabble in Egyptology.[9]

William Rivers, his colleague at St John's, changed all that. In the winter of 1900–1 he wrote from Upper Egypt, where he was investigating the colour vision of the labourers on the dig being conducted by Anthony Wilkin, urging Elliot Smith to take advantage of the opportunities Egypt offered – was he aware that the brains of pre-dynastic people were still preserved in the desert? Elliot Smith decided to see for himself. The first Egyptian grave he examined contained the skeleton of a boy with a stone in the bladder that was several millennia older than any other recorded. His attention was at once aroused; soon he was being invited by archaeologists to examine the bodies they were bringing out.[10]

'The prehistoric material is really most extraordinary, not to say marvellous', Elliot Smith wrote in November 1902. 'In some bodies the whole of the soft parts are retained in a desiccated state, and I have a large number of excellent brains (about 7,000 years old!) hair, beard; even eyes, nerves, muscles, genital organs and various viscera. There is nothing Negroid about the folk; but there are numerous resemblances to the Iberian type.' His involvement with archaeology quickly expanded when he was called in to settle a dispute as to how old the pharaoh Thotmes IV had been at the time of his death. Elliot Smith and the archaeologist Howard Carter resolved the matter by taking the mummified pharaoh in a taxi to a Cairo nursing home, where the only X-ray system in Egypt was in operation: they soon established that Thotmes had died at the age of about twenty-six. At Carter's suggestion, the director of antiquities, Gaston Maspero, entrusted Elliot Smith with the task of examining all the royal mummies then in the Cairo Museum. There were fifty in all, but only twenty had been unwrapped, so Elliot Smith set to work on the remaining thirty. By 1912, when he produced his classic work *The Royal*

Mummies, he was an acknowledged expert on the complex techniques of embalming and mummification.[11]

As an anatomist, however, Elliot Smith found himself constantly being involved in disputes between rival schools of archaeologists. When the British had arrived in Cairo in 1881, Lord Cromer, the proconsul who effectively ruled the country, had been in favour of leaving Egyptology to the French in return for concessions elsewhere: the occupation of Egypt by the British should not, he believed, be 'made an excuse for filching antiquities from the country'. In practice, though, the English had taken over and, by the time Elliot Smith arrived, the cast of the drama was well established. Ernest Wallis Budge, of the British Museum, would sweep in every year for a month or two, hoovering up artefacts and sending them back to London, in clear defiance of the restrictions on exporting antiquities; the Rev. Archibald Sayce, an elderly Oxford linguist, with his well-furnished Nile boat to accommodate his travelling library, offered tea and expertise to visiting Egyptologists; and, above all, Flinders Petrie, who had dug in Egypt almost every winter for forty years, living rough in tents and eating food out of cans, had revolutionised the technique of excavation, and trained many Egyptologists and archaeological labourers.[12]

Elliot Smith chose instead to align himself with the American archaeologist George A. Reisner, who had quickly established himself as the new force in Egyptian field archaeology, setting fresh standards of thoroughness with his excavations. It was through this connection that Elliot Smith took another decisive step. In 1907, the Egyptian government decided to raise the level of the Aswan dam by seven metres, thus flooding a huge area of antiquities. A vast archaeological rescue programme had to be organised and Reisner was put in charge. Elliot Smith agreed to be his anatomical adviser. The area was honeycombed with ancient cemeteries and he soon found himself confronted with an enormous task: when he arrived in Nubia in October 1907, some 2,000 burials (the results of less than a month's excavation) were awaiting him. In desperation he asked London for an assistant, and the Royal College of Surgeons sent him Frederic Wood Jones.[13]

It was the start of a lifelong friendship, albeit one that would be clouded by jealousy and rivalry. Wood Jones was then twenty-eight, the son of a Hackney builder who had risen in the world, and a protégé of Elliot Smith's fellow anatomist, Arthur Keith. He was working as a demonstrator at the London Hospital when the summons came. The two men met up in Cairo and then travelled south together. Having only encountered Elliot Smith's pontifical manner when lecturing, Wood Jones was relieved to discover that he was good company:

> The heat, the discomfort, the dust and the interminable delays seemed powerless to disturb him; he merely grunted when things were particularly trying and smiled when they were not. We were met at Shellal by a dust storm and, when that was done with, we started on our seemingly impossible task of dealing with thousands of waiting skeletons.

They had already agreed on the division of labour: Elliot Smith would measure and dictate the notes, Wood Jones would do the recording. They now set to work on some predynastic graves among the mud huts of an Egyptian village. Elliot Smith, his new assistant recalled,

> sat at one end of the grave and I at the other. The heat was terrific; the metal callipers became too hot to hold with comfort unless care was taken to place them in what little shade was thrown by one's own body. The flies were legion. They swarmed about our faces and crept into our eyes. Every now and then a hot swirl of dust, of very evil origin, would come circling across the dry mud plain and over the grave where we sat.

In his deep and level voice, Elliot Smith monotonously recited endless measurements and figures; Wood Jones was soon exhausted:

> I brushed the dirt and flies from my face and wrote, seeming without end, 'minimum frontal breadth ninety-one, bizygomatic breadth one thirty-seven, cranial breadth'. At intervals, I would look at my entirely serene and benevolent-looking chief and wonder how much longer I could stand it. It started again, thank God he had got to the leg, 'maximum length four thirty-seven, oblique length . . .' and then,

throwing the hot callipers into the sand, he gave his opinion of the dust, the heat, the village of Shellal and the ancient dead in one comprehensive and highly descriptive sentence that must have come straight from the remembered repertoire of his student days in Sydney.[14]

They called it off for the day and went to the suk at Aswan and bought cheroots that were so foul that even the Nubian flies kept their distance.

The anatomists' primary task was simply to record the sex of the skeletons, which was not as simple as they first thought. They also made some efforts to record the diseases among the bodies and the cause of death, becoming in effect pathologists: on one occasion they found the bodies of sixty-two men who had been executed, some by the sword and some by hanging, huddled together in a trench. And, above all, Wood Jones later remembered, 'it was not enough that a record be made of the anthropological measurements for future reference. The archaeologists needed to know the racial characters of the skeletons in graves as they were opened up.'

At the end of the season it was decided to return to Cairo on board a government steamboat. The Nile was low and the boat kept running aground on sandbanks, so the journey took three weeks. It was the perfect opportunity to write up all the notes the two anatomists had kept. But Elliot Smith refused to embark on the task and messed about like a lovable schoolboy. His thoughts seemed to be elsewhere. Wood Jones suspected that he was cooking something up; that some great theory was forming in his mind.[15]

Wood Jones thought the period in Egypt changed and enlarged Elliot Smith's outlook, but not 'because the romance of the land of the pharaohs attracted him or had him under a spell'; in fact he was remarkably oblivious to his surroundings. It was simply that the place furnished him with skulls and skeletons and mummies. Others disagreed. 'Egypt's spell ensnared him, to his own lifelong enthralment', a fellow anatomist believed.[16]

There is an element of truth in both views. His time in Cairo changed Elliot Smith in two important ways. Firstly, it transformed him into a major figure in anatomical history. Having previously charted the development of the brain in primitive pouched and egg-laying marsupials and in mammals and reptiles, he could now work extensively on human brains – not just ancient but also modern ones. 'In the autopsy room of the hospital [in Cairo] he found a wealth of opportunities to study the human brain', one of his students later explained. 'Whereas [others] from paucity of material, had to devote their attention to the slow processes of histology, Elliot Smith, with hundreds of brains at his disposal every year, was able to cover a vastly greater material by simplified methods of gross dissection, and this produced the convincing evidence till then lacking but absolutely essential to the elucidation of human brain structure as a basis for the study of function.'[17]

In 1904 Elliot Smith published a classic paper on the occipital region of the human brain, and two years later he travelled all the way to Belfast to read a paper on the visual cortex to the Anatomical Society. His journey was not wasted. 'I think the event of the meeting has been the rise of Elliot Smith, in the estimation of all the anatomists present, to the very first place among the men who work at our subject', wrote one of those who were present. A year later, Elliot Smith was made a fellow of the Royal Society, the citation singling out 'his researches in the morphology of the brain, which have thrown light on many obscure problems of cerebral organisation, and are amongst the most noteworthy contributions, in recent years, to verte-brate morphology'. In 1909 he was appointed to the chair of anatomy at Manchester University.[18]

The Egyptian period enabled Elliot Smith to complete the final panel in his great triptych of brain research – marsupials, mammals and reptiles, humans – and then to move on to the major questions which had always interested him. What part did the brain play in the evolution of life on earth, from the simple organ of the platypus to the sophistication of the human brain? What, for example, is the significance of the fact that the visual area of the human brain is set well back in the skull? Very soon he was coming up with some bold and original hypotheses.

In 1912 Elliot Smith offered a 'tentative sketch of the genealogy of man' to the British Association for the Advancement of Science. Although much remained to be discovered there was, he believed, one unquestionable category of evidence that the scientist could seize hold of and examine – 'the steady and uniform development of the brain along a well-defined course throughout the primates right up to Man'.

There was, for example, the development of the part of the brain known as the neopallium (now called the neocortex), an organ both of associative memory and of rapid action, which had enabled mammals to be adaptable and capable of learning directly from experience. Similarly, as mammals moved first into the trees and then on to the land, so their brains had developed. Originally the brain was dominated by smell, but once mammals began to live in trees, there was enormous development of the visual cortex and a reduction of olfactory parts of the brain. Arboreal life also gave importance to the senses of hearing and of touch. 'The process of devising the complex machinery for controlling skilled movements leads to the progressive development of the prefrontal territory,' Elliot Smith wrote, 'the size of which becomes a distinctive feature of the primates and in particular of monkeys, apes and man. The prefrontal area is intimately concerned not only with the acquisition of skill but also with controlling the automatic movements of the eyes, which are an essential part in fixing the gaze and concentrating the gaze upon some objects.'[19]

When it came to the crucial question of man's development from the apes, however, Elliot Smith believed that there simply was not enough evidence to reach firm conclusions. 'It was not the adoption of the erect attitude that made man from an Ape, but the gradual perfecting of the brain, and the slow upbuilding of the mental structure, of which erectness of carriage is one of the incidental manifestations.'[20]

In his own specialist area of anatomy, Elliot Smith showed mastery of the material, care in reaching conclusions and forming hypotheses, and great clarity and lucidity in exposition. But he was much less temperate in the other areas of academic debate with which he now got involved. Indeed, the second effect of his time in Egypt was to complete his transformation from a shy, anonymous, anatomist into an intellectual prizefighter with a considerable ego. How this happened remains something of a mystery. Perhaps it was simply the effect of associating with archaeologists, a notoriously combative and disputatious academic tribe; perhaps it was hubris, the cumulative effect of having known only success for so long; and, perhaps, mixed in with this, a colonial contempt for the high priests of orthodoxy in the Old Country.[21]

Elliot Smith's new, more dogmatic, persona was made apparent in

1910, when *Nature* reported that Flinders Petrie had discovered a stone tomb near the pyramid of Sneferu containing a body which had been 'completely unfleshed' before being wrapped in linen – thus proving, said the journal, that 'the dissevering of the skeleton was the custom among the highest classes at the beginning of the Pyramid period'. Elliot Smith, who had already had public disagreements with Petrie on burial customs, lashed out: 'All who are interested in the serious attempts that are being made to reconstruct the real history of ancient Egypt and to sift established truth from wild conjecture must deplore Prof. Flinders Petrie's attempt to revivify the corpse of a belief in the supposition that the archaic Egyptians were in the habit of cutting up their dead, a view that has been so effectively hanged, drawn and quartered during the last ten years.' Regardless of the issues, the tone caused offence: Elliot Smith and Petrie had a brisk private exchange.[22]

That was only the beginning. The following year, Elliot Smith produced the theory he seemed to have been forming on the Nile while returning with Wood Jones from the Nubian cemeteries. Examination of skeletons there had convinced him that at the beginning of the third millennium BC there had been an invasion of people from the north with heads somewhat rounder than those of the predynastic inhabitants of the country, and with features of the skull which at the time were known as 'Armenoid' or Eastern European. Elliot Smith believed that the pyramids and other monuments associated with ancient Egypt were in fact not the work of 'African' Egyptians but of these immigrants from the north, with their bigger brains.[23]

In a slim volume entitled *The Ancient Egyptians and their Influence upon the Civilization of Europe*, he forged a bold hypothesis from his own anatomical work and the American archaeologist Reisner's recent account of the origins of metalworking: that the skills of building in stone and working with metals had been developed in Egypt and then spread round the Mediterranean and from there outwards across Europe. Egypt was thus the crucible of European civilisation.[24]

While the manuscript was being completed, in May 1911 Elliot Smith visited Cambridge as a medical examiner. There he found students being tested on 'skulls with definite Armenoid characters', which turned out to have belonged to natives of the Chatham Islands, south of New Zealand, 'a place almost as remote from Egypt or Armenia as could well be found'. This discovery set him thinking about the movements of early man about the earth. When he called on William Rivers, his old friend at St John's College, then writing an enormous *History of Melanesian Society*, he found that Rivers had similar preoccupations. Indeed, Rivers had formed his own theory which he was planning to make public at a scientific meeting later that year. He knew that there would be strong opposition.

The two men agreed to pool their forces.[25]

Rivers in the 1900s

William Rivers had returned to Cambridge from the Torres Straits a divided soul, torn between his old allegiance to psychology and neurology and his new passion for anthropology. For the next decade, he tried to do both, combining teaching in Cambridge with fieldwork on the other side of the world.

Rivers embarked on several experimental projects in psychology. The most important, which had a lasting influence on his later work, was carried out with the lovable and eccentric neurologist Henry Head, with whom Rivers had worked at the National Hospital in Bloomsbury in the 1890s. After failing to get taken on to the permanent staff at Queen Square, Head had independently pursued research into the role of the nervous system in pain, carrying out detailed observations on patients while working in a series of hospitals in London and Liverpool. This pioneering work in a complex and neglected area of neurology – which would be ethically unacceptable today because it was done without the 'informed consent' of his patients – led to his election to the Royal Society at the age of thirty-eight. By 1900, Head was a consultant at the London Hospital, spending his time oscillating between private practice in Harley Street and the patients he saw in Whitechapel. But he had become increasingly frustrated by the difficulty in fitting research into his life as a busy clinician and irritated by the unreliability of patients in reporting their symptoms. He therefore decided to experiment on himself and enlisted Rivers' help. The object was to establish how injury to the nerves affected their functioning.[1]

Between 1903 and 1907 Head spent some 170 hours in Rivers' room at St John's College in Cambridge, observing the recovery of sensation in his arm after the nerves in it had been severed. The two sat at a large table with Head's arm exposed between them, Head with eyes closed while Rivers touched or pricked the skin in various ways recording his own stimulus and Head's answers. Otherwise they did not speak. Head wrote to his wife Ruth in November 1903:

> With my eyes closed I try to let my thoughts flow by like clouds on a windy day. No one thought must occupy attention permanently and I must entirely detach myself from the idea that experiments are in progress. Suddenly in this flowing sea of thoughts there appears a flash of pain a wave of cold or the flicker of heat – It should appear with the suddenness of a porpoise, attract attention and again disappear leaving the untroubled sea to its onward flow. Such is the most perfect condition for psychological investigation: this state I can now assume at will.

After a while Head would open his eyes and have a short rest, while dictating an account of what he had recently experienced. After a further pause, the procedure was resumed, but they found that it was only possible to work for about an hour at a time. To relieve the tension, the two men had interludes of vigorous physical activity. 'Rivers and I ran up and down the John's wilderness like boys', Head wrote to his wife. In the summer they boated on the Cam and on one occasion, after a 'good morning's work on my arm', took horse and rode for many hours into the countryside.[2]

As the experiment proceeded, Head's will was tested, but he was determined to continue. 'The arm is beginning to recover and the returning sensation is most aggravatingly difficult to test', he wrote in July 1903. 'I prick a portion of skin with a needle or pull a hair and at the end of an hour a number of points are found and marked where pain can now be felt. After suitable rest Rivers tests these places and finds that all the sensitive spots have now become insensitive and the only places where I can feel are unmarked. It is trying wearisome work and all the time we are haunted by a fear of forgetting to test some important point until the suitable period for such testing is past. I shall know a great deal about pain by the time this experiment is over.' Head's wife was alarmed but his mother had other concerns: 'I am not half as miserable when I think of what Harry has done to himself as when I think of the poor dogs and animals who are tortured without the chance of refusing.'[3]

Head and Rivers eventually concluded that the nervous system contained two separate layers – one powerful but very crude, the second more localised and subtle – which they named protopathic and epicritic. The two systems, they wrote, 'owed their origin to the developmental history of the nervous system. They reveal the means by which an imperfect organism has struggled towards improved functions and physical unity.' But this dichotomy had a wider significance: 'The gradual triumph of the epicritic over the protopathic in human evolution could be seen as a metaphor for the triumph of civilisation over savagery in human history.'[4]

We may wonder how important this experiment was to Rivers. After all, Head was the moving spirit. Yet Rivers' pupil, Frederic Bartlett, thought that it defined the rest of his academic career. He argued that, although Rivers began and finished his academic life as a psychologist, 'his preferred habit of thought was that of a practical logician'.

> He liked to adopt, or find for himself, some general principle, then to search untiringly, but with complete fairness, for illustrative material and finally, in the light of this material to regard the basic principle as established.

Bartlett argued that practically everything of length that Rivers wrote, whether sociological, anthropological or analytical, followed closely the general theory that he and Head had arrived at. There would be a basic primitive organisation, little differentiated and subject to an 'all or none' type of expression, which would then be invaded by incoming elements that appeared to be integrated with the primitive organisation. The new elements might appear to transform or even to supplant the original organisation but when it came under strain, the old primitive element would reveal itself again.[5]

Some modern scholars have seized on the fact that Head's glans penis featured in the experiment to suggest that the two men were having a homosexual relationship, but there is no evidence to support this theory. Head had been happily married since 1900 to Ruth Mayhew, the bluestocking daughter of the chaplain of Wadham College, Oxford, who shared his intellectual interests and to whom he corresponded intimately and constantly when away from home. On Rivers' side,

there may perhaps have been some homoerotic element in the relationship, but – as would later become clear – he was not a man with
strong sexual feelings and such as he had he repressed. Far more
important to him would have been the pleasure of intellectual
comradeship with Head. Himself intensely cautious intellectually,
Rivers liked to be with other men who were prepared to be bold and
adventurous.[6]

Certainly, when Rivers operated on his own, as in the other major
psychological project he conducted in this period, caution was well
to the fore. The series of experiments on the effects of alcohol on
fatigue which he described in lectures to the Royal College of
Physicians in 1906 were notable more for their thorough methodology
than for any bold conclusions. He insisted on 'the absolute necessity
that experiments with drugs on man should be carried out with
adequate control, designed to exclude the influence of interest, sensory
stimulation and suggestion'; some scholars claim that this was the
first use of controls in medical research trials. Rivers carried this
principle to the lengths of himself abstaining from all alcohol, tobacco,
tea and cocoa for two years. In the end, his conclusions were mainly
negative: that the findings of previous researchers should be ignored
because their methods had not been wholly scientific.[7]

But Rivers' real passion at this time was anthropology. In the long
summer vacations, he travelled to the other side of the world, to apply
the genealogical method of enquiry he had developed in the Torres
Straits, which he believed provided an important new tool, a means
of quickly extracting concrete information from tribesmen who were
completely bemused by abstract western ideas, such as 'government',
yet often had detailed and precise knowledge of family relationships
stretching over many generations.

In 1901 and 1902 Rivers spent five months among the Todas, a
pastoral tribe living in the Nilgiri Hills, in the south-west of India. As
a community of some 800 people, they provided an ideal subject for
the genealogical method; but they had also attracted considerable
attention already because they practised adelphic polyandry (a wife
married her husband's brothers as well as her husband) and their
strikingly Semitic appearance had prompted speculation that they
were 'God's ancient people' or the remains of some ancient Roman
colony.[8]

What struck Rivers most about the Todas' daily life, however, was the prominent place occupied in it by the buffalo. The care and culture of their buffaloes were associated with much ceremonial, and formed the basis of the greater part of their religious ritual. With the Todas a priest was a man especially set apart for attending these buffaloes, and their temples were the dairies in which the milk of the sacred animals was churned. Every operation of the dairy thus became a religious act, and every incident in the lives of the buffaloes a pretext for religious ceremonial.[9]

Living in a bungalow and operating through two interpreters (who were both Christians) and with some help from a local colonial official, Rivers' fieldwork was not without its difficulties. At one stage, just when 'he had hoped to overcome the scruples of the people and to obtain information on many doubtful points', three of his main informants on sacred matters each suffered calamities; and Toda diviners decided that the gods were angry 'because their secrets had been released to a stranger'. As a result, his sources of information 'ran dry to a certain extent'. Similarly, Rivers' attempts to make sense of certain of the Todas' practices proved fruitless – 'the Todas were quite unable to give explanations of their customs', he wrote later, 'the answer to nearly every question being that the custom in question was "ordained by the goddess Teikirzi"'. As a result, Rivers was forced to provide his own explanations for the motives behind the Todas' behaviour, and was not always able to do so: 'I could not satisfy myself as to what the people really had in their own minds when they were lamenting' a dying buffalo.

Similarly Rivers never found out why the Todas combined the strictest regulation of marriage with what he regarded as almost complete sexual promiscuity both before and after marriage. He remarked that their religion was more sophisticated than might be expected 'for a people living in such simple circumstances', but while their idea of 'god' was highly developed, he saw a certain 'degeneration' in their rituals involving bells and dairy vessels. His tentative explanation was that the Todas had migrated a thousand years ago from the Malabar region, where they had absorbed Christian and Jewish influences, to the Nilgiri Hills, where, in the words of one scholar, 'under special environmental conditions their religion had undergone simultaneously a theological degeneration and an overelaboration of ritual'.[10]

An anthropologist who visited the Todas thirty-five years later 'found little that Rivers had missed, and little on which correction was needed'. There was one glaring omission, however, which calls into question the adequacy of his genealogical method. Rivers had failed to notice 'the existence of a system of matrilineal clans, in addition to the more obvious patrilineal ones in which the sacred dairies of the Todas, their most fundamental institution, are vested'.[11]

Nevertheless Rivers managed to assemble enough information to produce an immense book, published in 1906. One reviewer commended *The Todas* for the 'astonishing mass of material that it places at our disposal', singling out as an example of Rivers' thoroughness 'the 72 genealogical tables of Toda families, comprising almost all of the 800 people that compose the tribe, which form the basis of his explanation of family law, and which are in themselves one of the most valuable documents ever furnished to ethnology'. There were a few dissenting voices who felt that Rivers never engaged emotionally with the Todas and remained an outsider; others thought that the book failed in literary terms, losing its way in a mass of detail, and were disappointed by the failure to draw conclusions. But, overall, *The Todas* was a triumph which at once established Rivers as the leading British anthropologist and seemed to put his genealogical method firmly on the map.[12]

By the time he carried out his next extended field trip, in the winter of 1907, however, Rivers' thinking had advanced considerably. He had become aware that, far from inventing the genealogical method, he was in fact only the latest in a long line of anthropologists to use it; indeed he realised that he had wandered on to jealously guarded territory. 'Kinship', as the subject was usually known, was a field beset with controversy and disagreement, dominated by Americans who had worked with North American Indians and tended to assume that their findings held true for all 'primitive' peoples. Just as Rivers had earlier revived and championed Gladstone's theories on colour vision, now, in approaching kinship, he resuscitated the older ideas of the American scholar, Lewis Henry Morgan. Writing in the 1870s, Morgan had believed that it was possible to reconstruct the history of primitive people from the terms they currently used to describe their patterns of marriage and social organisation, back to prehistoric times when they had consisted of sexually promiscuous human bands.

Although Morgan's theories had by the 1900s attracted much criticism, Rivers believed that they still offered an important way of investigating human society. 'My previous work in Torres Straits and India had shown me the fundamental importance to the science of Sociology of the method of counting relationships which among most of the races of mankind differs so greatly from that practised by ourselves', he wrote. 'I had reached the belief that in systems of relationship we have, like fossils, the hidden indications of ancient social institutions and that their study is essential for advance in our knowledge of prehistoric sociology.'[13]

His trip in 1907 was therefore designed to make a sustained assault on patterns of kinship in the Pacific. After spending two months in Hawaii and Fiji, testing Morgan's theories, Rivers travelled on the Melanesian mission's schooner *Southern Cross*, a modern ship of 590 tons, capable of doing ten knots which had enabled the missionaries to visit islands they had not been to before. The ship had a decent library, chapel and quarters for women, and was commanded by the redoubtable (and unfortunately named) Captain William Sinker. Rivers visited the New Hebrides, the Banks, Torres, Santa Cruz Islands, and finally the Solomon Islands, where he conducted intensive fieldwork with Maurice Hocart, a 25-year-old protégé of William McDougall. Six months later he revisited the same islands in reverse order.[14]

This was survey, not intensive anthropology; a throwback to the methods of earlier times – if not armchair anthropology, it was certainly deckchair anthropology. Rivers admitted that 'much of the work was collected during hasty visits to islands, sometimes of only a few hours' duration', this information being further supplemented by 'work with natives who were travelling from one island to another'. He also conceded that 'much of it can only be taken as a rough preliminary account of social conditions which will, I hope, be more thoroughly studied before it is too late'.[15]

This was the happiest period of Rivers' life. Away from the pressures of Cambridge, spending time in the enjoyable company of sailors, savages and missionaries; living the outdoor life, but in some comfort, with his health under control; pursuing work he found intellectually absorbing: all the elements which made him happy were present. As the historian Ian Langham has pointed out, 'Time and again in his later writings, and especially when he is waxing autobiographical, Rivers refers to incidents which took place during or shortly after the 1908 expedition.'[16]

He and Hocart got on very well together. Hocart was the kind of reserved, highly intelligent young man Rivers always liked, formidably grounded in classical languages, comparative philology and psychology. Was there more than intellectual comradeship to the relationship? A reference in a letter which Hocart sent to Rivers a few years later, to having 'flute duets' on Fiji with a 'precocious dandy' has led Ian Langham to suggest that the two men may have shared some sort of romantic idyll with their servants while in the Solomons. This, from all we know of Rivers, is extremely unlikely.[17]

What is true is that Rivers established a greater affinity with the Solomon Islanders than he had with the Todas or on Murray Island. He and Hocart accompanied the 'native medicos' on their visits to their patients, noticing that they used techniques of massage very similar to those of a European therapist; they were also able to gather intimate details on the practice of circumcision and, on Eddystone Island, gained a full account of sexual relations and marriage. 'Soon after puberty defloration of a more or less ceremonial character takes place,' Rivers noted, 'and this is followed by a period in which a girl was at the service of any man on payment of a fee to her parents. Once married, however, the rule is absolutely strict, though no doubt it is sometimes broken, that neither man nor woman may seek other partners.'[18]

But what really excited Rivers was the chance to penetrate much more deeply into the islanders' systems of belief and understanding of the world. On Eddystone Island, for example, where he and Hocart worked together for a month, they obtained a long account of the destination of man after death. 'We were told that he stays in the neighbourhood of the place where he died for a certain time until the spirits arrive in their canoes from a distant island inhabited by the dead to fetch the ghost to his new home', Rivers reported. They were even able to observe this phenomenon for themselves. 'On one occasion we were present in a house packed tightly with people, who heard the swish of the paddles of the ghostly visitors, and the sound of their footsteps as they landed on the beach, while for several hours the house was filled with strange whistling sounds, which all around us firmly believed to be the voices of the ghostly visitors come to fetch the man who had lately died.'[19]

While Rivers did not himself hear the swish of the paddles, he understood that for those in the hut it was entirely real. His writing of this period has a new tone of respect for the logic of savage belief systems; in modern jargon, it reveals cultural relativism. It seems that his faith in the superiority of the higher races over the lower has faltered a little. One day, when Rivers was travelling in a boat with four inhabitants of Niue (or Savage Island), he took the opportunity of questioning them about their social organisation. They answered fully but then, to his surprise, asked if they could now question him about *his* customs.

> Using my own concrete methods, one of the first questions was directed to discover what I should do with a sovereign if I earned one. In response to my somewhat lame answers, they asked me point-blank whether I should share it with my parents and brothers and sisters. When I replied that I would not usually, and certainly not necessarily do so, and that it was not our general custom, they found my reply so amusing that it was long before they left off laughing.[20]

The broader geographical scope of the 1907–8 expedition – travelling by boat and visiting a number of islands – provided Rivers with a perspective different from his earlier work. His agenda began to move away from questions of social organisation within a single community towards wider comparative issues and matters of history. Why was one island different from another? Why were some customs found in some places and not in others? This led Rivers towards questions of cultural influence and historical colonisation but he only really came to realise this when he began to write up his material, back in Cambridge.

A glimpse of Rivers at this time has survived. In 1909 a young psychology student at St John's received an illegibly written note asking him to call on Dr Rivers for tea one afternoon:

> At Cambridge you knock and enter. The room was beautiful, with its brown panelled rooms; but nothing else was. It was in an awful muddle, with books and papers and odds and ends of anthropological trophies all over the place . . . Then Rivers came out of his inner study, and somehow at once the room came alive, and the things in it were right

after all. There he was, rather tall, trim, quick and light in his movements, in navy blue. You got a swift impression of straight, broad
shoulders and a jutting chin, and at once of a tremendously alert mind.
He shook hands, told me to sit down, sat down himself, said that no
doubt I knew what I wanted and how to get it, took off his spectacles,
swept his hand across his eyes with perhaps the most familiar of his
gestures, and waited . . . He was like a man suddenly come back from
somewhere into a world which on the whole he did not like very much.
Tea came up from the College Kitchen. He poured me out a very
strong cup, snorted audibly when I had little milk and no sugar, and
contented himself with sweetened milk and water. I had a bit or two
of rather stale bread and butter and a piece of Madeira cake that might
almost as well have been made of sawdust.[21]

The young student, Frederic Bartlett, would in time become the
greatest psychologist Cambridge ever produced, but he came from a
humble background in Gloucestershire and his training was in philosophy, so he was a little out of his depth.

Somehow it came out that I had read a little anthropology, and even
that I had heard of Cross-cousin marriage and Classificatory System.
Rivers' stammer disappeared. The table was cleared of a book or two.
For a brief time we pored over complicated diagrams of relationship.
Only for a short time: the History of Melanesian Society was urgent, and
out I came again, suddenly to realise that I had been treated, not as
an undergraduate, but as an equal.
    That was Rivers' way, then and later. It was a great part of his power
over men, especially younger men.

Bartlett, who found the whole business of kinship systems very
daunting, was hugely impressed by Rivers' ability to remember all the
complex details of relationships in his Melanesian material. In fact,
though, Rivers was struggling to make sense of what he had gleaned
in the South Seas. In an unpublished fragment he later recalled how:

One May term I was engaged on trying to discover the meaning of a
number of peculiar forms of marriage which occur or were formerly
practised in Melanesia. The whole set of marriages with their

accompanying conditions formed a confused tangle of facts and processes, the meaning of which I wholly failed to see. Morning after morning I awoke with my head full of their complexities and the days were spent in . . . work by means of which little features of one island were brought into relation with those of another, and the scheme became clear, but there still failed completely to be any central thought which would bind them all together. At length I got into such a state that I hardly slept at all. I woke so early so often that my short hours of sleep were having a serious effect on my health.

It is very rare to get this kind of glimpse of Rivers' thought processes. His solution to the crisis was characteristic:

I decided to go home to Ramsgate for a few days and following my usual custom of never travelling by land when it is possible to go by sea, I went to London one evening, slept the night there and took the river steamer . . . in the morning. The journey had the favourable effect on me that sea or river [voyaging] always has in allowing me to live without thinking and I had a quiet and more or less somnolent day with the minimum of mental activity of any kind. I went to bed very early and woke the next morning with the solution to my difficulty . . . The solution in question was that various forms of marriage were the results of a former state of dominance of the old men and monopoly of the younger women.[22]

These 'anomalous' marriages – in some cases grandfathers were marrying women of their grandchildren's generation – could only be explained if a state of 'gerontocracy', or domination of the society by the elders, had once been widespread in Melanesia. The likelihood was that it had been introduced from outside.

This discovery, when combined with a linguistic comparison of the actual kinship terms in use and other forms of evidence, such as the existence of different hut designs and the survival of secret societies on the islands, forced Rivers to have a complete rethink: Melanesian culture could only be explained as being the product of a series of invasions. What he had witnessed, Rivers began to realise, was a culture made up of several different strands interwoven over time. Many of the social practices of the Melanesians appeared to have been adapted from other cultures, brought by immigrants. 'An isolated people do not invent or advance, but . . . the introduction of new ideas, new instruments, and new techniques leads to a definite process of evolution, the products of which may differ greatly from either the indigenous or immigrant constituents, the result of the interaction thus resembling a chemical compound rather than a physical mixture.' This suggested that the customs and culture which Rivers had witnessed in Melanesia in the 1900s had not necessarily been created by the inhabitants themselves. They might have been brought there from afar, over the seas, by immigrants; who might not necessarily have stayed.

It was at this time, May 1911, when Rivers' thoughts had reached this stage, that Grafton Elliot Smith turned up in Cambridge, having just corrected the proofs of *The Ancient Egyptians*. As we have seen, it was in Cambridge that he noticed that the skulls which medical students were examining had an 'Armenoid' appearance, despite coming from Chatham Island in Oceania. And it was then that he and Rivers began to compare notes.[23]

One might think that an anatomist with a sideline in Egyptian history and a psychologist with an interest in the social systems of Oceania would have nothing in common. Yet both had, from different perspectives, begun to 'entertain the idea of extensive movements of early man about the earth'. Initially, it was the Australian who had doubts and the Englishman who urged him forward. 'Your remarks in June quite converted me to your point of view', Elliot Smith wrote later in the year. 'Knowing nothing of the evidence on your side when I wrote the book, I naturally took the more cautious line that the culture may have spread amongst kindred peoples without any great racial movement . . . In the light of your work, I have no hesitation whatever in speaking now of this movement as the explanation of the spread of culture.'[24]

In September 1911, in his presidential address to the anthropology section of the British Association, Rivers announced a fundamental shift in his thinking – his conversion to the idea that in the prehistoric world, culture had been spread by conquest and migration and had not emerged independently. This was contentious stuff, as Rivers acknowledged, because British anthropology was primarily inspired by the idea of evolution founded on a psychology common to mankind as a whole, and, further, a psychology differing in no way from that of civilised man. German anthropology, on the other hand, focused on the mixing of cultures and races. 'I have been led quite independently to much the same general position as that of the German school. With no knowledge of the work of this school I was led by my facts to see how much, in the past, I had myself ignored considerations arising from racial mixture and the blending of culture.' He then described at some length how the study of social structure in Melanesia had driven him to this conclusion. At the same meeting, Elliot Smith developed further his ideas about the influence of the Egyptians on early European civilisation.[25]

The two men were now launched on a common enterprise, in which they coordinated and supplemented each other's work. In 1914, Elliot Smith christened the new doctrine 'diffusion', and 'diffusionism' is the label that historians now attach to their work. Elliot Smith's next step was to argue that the dolmens, or stone megaliths, of the Mediterranean were crude copies of Egyptian mastabas or rock tombs and, thence, that the spread of megalithic monuments around the world – to the British Isles to the west, to Japan and America to the east – was due to the influence, directly or indirectly, of Egyptian civilisation. Then, as the reigning authority on Egyptian royal mummies, he turned to the study of burial practices around the world. In the meantime, Rivers, though still preoccupied by his Melanesian project, investigated another problem. How could it have been possible for the peoples of Oceania to have travelled vast distances across the Pacific when their modern-day descendants scarcely ventured away from the shore? The answer, Rivers argued, was that it was quite common for skills to fall into disuse when they were no longer required; the fact that these people no longer had the art of navigation did not mean they had not possessed it in the past.[26]

For the next decade Rivers and Elliot Smith were close intellectual allies. The fact that Rivers chose to ride in harness with Elliot Smith, and to lend his enormous prestige to the latter's ideas, infuriated some of his friends. Alfred Haddon, for example, who wrote a history of human migration and believed the diffusion of culture to be an important element in world history, nonetheless felt that Elliot Smith was blowing it out of all proportion. What then bound Elliot Smith and Rivers together? The two men shared a common intellectual position and a slightly puritanical outlook. Elliot Smith was at this time a formidable figure, the coming man in British anatomy, with a forceful personality. Rivers, as his relationship with Henry Head showed, was inclined to be influenced by strong, intellectually dominant men. And, after a decade and a half spent in the cautious, thorough accumulation of information, he was finally ready to express himself. As one of his best interpreters has written, 'diffusionist speculation at last provided an outlet to his locked imagination, and he let himself go'.[27]

In July 1914, the annual meeting of the British Association for the Advancement of Science – which had been the main forum for scientific argument since the days of Huxley and Wilberforce – took place in Australia. Haddon, Rivers, Elliot Smith and Seligman all sailed out, taking with them a young Polish-born anthropologist, Bronisław Malinowski. Before the meeting proper, Elliot Smith visited the museum in Sydney and examined several mummified bodies collected from the Torres Straits in the 1870s by the Australian zoologist, William McLeay. He was surprised to find that the techniques used to preserve these bodies were exactly the same as those he had previously seen on Egyptian mummies. For example, 'the curious, inexplicable Egyptian procedure of removing the brain, which in Egypt was not attempted until the XVIIIth dynasty – i.e. until the embalmers had had seventeen centuries' experience of their remarkable craft – was also followed by the savages of Torres Straits!' He was convinced that these Papuan mummies 'supplied us with the most positive demonstration of the Egyptian origin of the methods employed'.[28]

Hastily amending his address to the conference, Elliot Smith announced his new findings triumphantly, clearly expecting to overcome the doubters in the audience. But his assessment of the Torres Straits mummy was not generally shared. The Oxford archaeologist J. N. L. Myres was quite unimpressed. 'What was more natural than that people should want to preserve their dead', he demanded. 'Or that, in doing so they should remove the more putrescible parts? Would not the flank be the natural place to choose for the purpose? Is it not a common practice for people to paint their dead with red-core?' He refused to accept that there was anything distinctively Egyptian about the techniques used. Alfred Haddon, who spoke next, was gentler in manner but equally unconvinced. The incision made to the feet and knees of the Torres Straits mummy were not suggested by Egyptian practice, and had been made simply to drain fluids from the body. Moreover, there were no other signs of communication between Egypt and Papua to explain how the custom of preserving the dead might have been spread.

Clearly it would take more evidence for the diffusionist case to prevail. Elliot Smith returned to England, vowing to find it.

PART FOUR

THE MOST EXCITING
DECADE

'The secrets of human nature'

At the end of the Torres Straits expedition, William McDougall took his time returning to England from Borneo, via China, Java and India. He knew it was now time to concentrate on one branch of work and, having decided not to take up anthropology, he returned to psychology – and to his ambitious, chosen task of a 'direct attack on the secrets of human nature'.[1]

How was this attack to be mounted? McDougall had a clear strategy. In September 1899, soon after returning to Cambridge, he arranged to give a course in experimental psychology at University College London, the following year. He would make use of the laboratory which James Sully, the professor of psychology there, had bought from the German psychologist, Hugo Münsterberg, when the latter took up William James's chair at Harvard. McDougall wrote to Münsterberg, asking for advice on how to prepare himself:

> You are, of course, aware that in this country exp. psych. is in a very backward state, that we have but few workers and very inferior laboratories. I am anxious to do something to remove this reproach from us and I mean to devote myself to the advancement of this branch of science.

Obviously his apprenticeship with Rivers in the Torres Straits had not been of much help. Enclosing a paper which had attempted to 'draw the two sciences, physiology and psychology, a little closer together', he added that he had been advised to visit the American schools of experimental psychology, and asked if he might work in Münsterberg's laboratory at Harvard the following spring. 'I must apologise for asking this favour of you,' the letter concluded, 'being as I am a complete stranger, but I have been encouraged to do so by reading your very interesting and stimulating little volume *Psychology and Life.*' But nothing was to come of it and McDougall did not go to Harvard, because his life suddenly became complicated.[2]

Cambridge tried to hang on to him. At the end of 1899, St John's College appointed McDougall to a tutorship, a post which carried a decent salary and involved several hours of weekly supervision in college. But on 31 May 1900 the Master of St John's received a flustered, embarrassed letter asking him to put into the hands of the College Council McDougall's resignation of the tutorship to which it had so recently appointed him. 'I feel compelled to do this', McDougall wrote, 'because, quite unexpectedly, I find myself about to marry, and I am conscious that the Council may have been influenced in making the appointment by the fact that I was a bachelor. I feel also that the duties of a tutor and a married man must be very difficult of performance by any one person.'[3]

In fact when McDougall wrote that letter, he was already married and living in Göttingen in Germany with the woman who was expecting his child.

When Annie Amelia Hickmore and William McDougall were married, in Kensington Registry Office on 23 May 1900, neither of their families was present. Her father's profession was given as 'government contractor' but on census returns between 1871 and 1901, Henry Hickmore, of 3 St Martin's Place, Brighton, was content simply with 'chimney sweep'.[4]

Annie, born in 1879, was twenty-one at the time of her marriage, eight years her husband's junior. It was her 'lovely smile and expression, at once human and divine' which captured his heart: 'I married almost suddenly for love pure and simple,' McDougall wrote years later, 'though I did not utterly lose sight of eugenic considerations.' She was not, to the modern eye, a great beauty: she did not have the face of a temptress, but of a plain, decent, working-class woman, with English, South-Saxon looks. But, for someone raised on Tennyson and the Pre-Raphaelites, with a strong racial sense, her long, lustrous hair would have added to her allure. How these two people from very different worlds met remains unclear, though there are many models of such liaisons across the classes in late-Victorian England

– Pre-Raphaelite painters who pulled girls off the street to sit as their models, or the Cambridge don A. L. Munby who was sexually aroused by mannish working women. It seems likeliest that William and Annie would have met in a train carriage or a public house, her origins being too humble to have allowed her to work as a hospital nurse.[5]

McDougall's autobiographical essay makes no reference to Annie's social class but alludes obliquely to the suddenness of the wedding:

> My marriage at the comparatively early age of twenty-nine was against my considered principles; for I held that a man whose chosen business in life was to develop to the utmost his intellectual powers should not marry before forty, if at all. But nature was too strong for principles; and I have never regretted the step.

Nor does he tell us anything of how they met, but conveys something of their social and intellectual disparity:

> It might be thought that for a charming young girl to marry an intellectual monstrosity like myself would be like making a bedfellow of a hedgehog. But my wife has proved equal to the task she undertook. In intellect and temperament we were as unlike as possible, pure complementarities: I introverted, reserved, outwardly cold and arrogant, severely disciplined, absorbed in abstruse intellectualities; she extroverted, all warmth and sympathy and charm and intuitive understanding.

One can only speculate about McDougall's deeper motives. Clearly he needed to marry; if he was to function as a loner, he had to have a companion, a helpmeet. Did he also subconsciously need a reason to escape from Cambridge, to remain outside the group? But how far, by marrying outside his social class, did he further enforce his isolation?[6]

McDougall must have moved fast to organise the trip to Germany. Very soon after their marriage, he and Annie arrived in Göttingen, where he carried out research on vision in the laboratory of the celebrated psychologist, G. E. Müller. The baby, a girl called Lesley, was born there in 1900 and, soon afterwards, the family returned to England and set up home in a charming house near Haslemere in

Surrey. Over the next decade and a half, four more children followed. McDougall was by all accounts a good husband and father. He certainly thought he was:

> To do one's duty by a wife and five children does require the expend-
> iture of considerable time and energy that might *possibly* be given to
> purely intellectual tasks. But I have always found delight and recreation
> in my home; I have never ceased to grow more grateful to my wife
> for her influence upon me and for her perfect exercise of the privileges
> of her position . . . she has saved me from entanglements which, if I
> had followed my principle, might well have wrecked me. Then, too, I
> have learned more psychology from her intuitive understanding of
> persons than from any, perhaps all, of the great authors. I venture to
> think that the success of our marriage has been partly due to my
> recognition that the intellectual is apt to ruin his domestic relations by
> permitting himself to regard them as of less importance than his work.
> At a very early stage I resolved to avoid that error.[7]

But this was not quite the whole story. Like many intellectuals, McDougall was embarrassed by a wife who was his social and intel-lectual inferior, so he shut her away in the country and tried to keep her from academic society. It seemed to one Cambridge observer that McDougall was 'tackling his marriage the wrong way'. 'We thought it was a mistake. He was making too much of it', Charles Myers' wife Edith told an interviewer many years later. 'He lost so many friends through that and I don't think he did himself any good by belittling her and not letting her mix . . . There was nothing against the woman. She was really very *nice* . . . And she was a good mother and a good wife. We thought that was all that mattered. They had a delightful family of quantities of boys and girls. And we were very sorry that we didn't see more of them.'[8]

Whatever embarrassment Annie may have caused McDougall in public, she undoubtedly gave him emotional security. This, and a modest private income, enabled him to be highly productive over the next few years, working mainly in a small laboratory he built in the attic at Haslemere and occasionally going to London to teach. Now that he had broken away from Cambridge – and was not going to America – McDougall took steps to repair the relationships which had broken down in the Torres Straits. He was one of the moving spirits behind the creation of the British Psychological Society in 1901, with William Rivers as another of the twelve founding members. Charles Myers, then convalescing in Egypt, joined the society soon afterwards and became its secretary in 1904, with McDougall as his deputy. All three served on the society's Committee of Management.[9]

As always, however, McDougall's social instinct had to compete with his urge to outdo his peers. His first research project after returning to England was an act of rebellion against Rivers. At Cambridge, Rivers had been an enthusiastic advocate of Ewald Hering's four-colour theory of vision, which he had picked up from his friend Henry Head, and on the voyage to the Torres Straits had got McDougall to read the proofs of an enormous chapter he had written, setting out Hering's theories. That alone would have been enough to provoke McDougall into revolt; the fact that this view was becoming orthodox only drove him further. 'Whenever I have found a theory widely accepted in the scientific world, and especially when it has acquired the nature of something of a dogma among scientists, I have found myself repelled into scepticism', he wrote later. In June 1901, he publicly dissented from a lecture on colour vision Rivers had given. 'I happen to be interested in maintaining a different view,' McDougall declared, 'all of the evidence from which this conclusion has been drawn is possibly capable of bearing a different interpretation.' He now devised a series of ingenious experiments designed to disprove Rivers' and Hering's theories on colour vision, and reinstate the rival three-colour theory of Thomas Young and Hermann von Helmholtz.[10]

In 1905 McDougall revealed his ambition with his first book, *Physiological Psychology*. He boldly used modern research into the workings of the brain and nervous system, such as Charles Sherrington's theories of synapse and reciprocal inhibition, to answer the questions about the mind posed by traditional, introspective psychology. Going far beyond the limits of then current knowledge, McDougall speculated on the physiological basis of consciousness and attention, on the nature of inhibitory process and on how drugs, fatigue and hypnosis acted on the human system. He was exploring the possibility of a physiological psychology – if only, ultimately, to point out its limitations. Bringing together current neurological concepts, the psychology of William James and some of the building blocks of

Darwinian theory, it was a remarkable piece of work for a 34-year-old and led to his being appointed to a readership in mental philosophy at Oxford.[11]

In some ways it was an ideal job for McDougall. The small salary supplemented his private income. His teaching load was light – only two weekly lectures for twenty-one weeks a year – and he had complete freedom to range at large over the whole field of psychology, which he proceeded to do. After teaching in both London and Oxford for some years, he built a large country house at Boar's Hill, a few miles west of Oxford, and installed his growing family in it.

Against that, however, he had to contend with Oxford's massive indifference. In the 1900s, Oxford was politically conservative and socially exclusive. The dominant colleges – Christ Church, Magdalen and New College – had a large percentage of Old Etonians, while many colleges did not admit non-white students. The drunkenness, prejudice and anti-intellectualism of the undergraduates were indulged by dons fearful of judging their social betters. In Magdalen it was bad form to pass the prelim examination at the first attempt and a Brasenose man who gained a first was ducked for 'breaking the traditions of the college'. Academically, Oxford was dominated by Greats, modern history and natural science. 'The School of Literae Humaniores', the 1913 *Student's Handbook* explained, 'is admitted on all hands to be the premier School in dignity and importance. It includes the greatest proportion of the ablest students, it covers the widest area of study, it probably makes the severest demands, both on examiner and candidates, it carries the most coveted distinction.' In Compton Mackenzie's novel *Sinister Street*, the hero Michael Fane remarks, 'It does not seem to me that one gains the quintessence of the University unless one reads Greats.'[12]

Psychology had no status at Oxford, no recognised place in the curriculum or examinations. McDougall was not even a member of the university because he was not attached to a college. 'If I had been recognised as a teacher of science, my punishment would have been light,' he wrote later, 'for by that date, science was well established in Oxford. But I was neither fish, flesh, nor fowl. I was neither a scientist nor a philosopher *pur sang*. I fell between two stools. The scientists suspected me of being a metaphysician; and the philosophers regarded me as representing an impossible and non-extant branch of science'. The readership in mental philosophy – 'that is, psychology camouflaged so as not to offend Oxford susceptibilities' – which McDougall held had been established by Henry Wilde, a retired industrialist who had an admiration for John Locke and did not believe that the workings of the mind could be studied scientifically at all.[13]

Whereas in Cambridge some of the philosophers had been sympathetic to the new discipline of psychology, in Oxford they were implacably hostile, denying not only the relevance but even the possibility of a science of psychology. The leading lights of the dominant neo-Hegelian idealist school argued that if psychology was anything it was descriptive philosophy and that Plato and Aristotle were the best psychologists; while modern psychology pretended to have made an advance towards a scientific understanding of the facts, all it had actually done was to create new names for them.[14]

McDougall made the best of it. The professor of physiology let him use several rooms to carry out psychological research in a private capacity and he gathered around him a few disciples, most of whom were reading Greats, and some of whom, such as William Brown and Cyril Burt, later became important psychologists. Burt's tutor allowed him to attend McDougall's lectures, but added that he would probably do better to spend his spare time on the river. '"Nothing," I was assured, "will ever be discovered about the mind which is not already to be found in the writings of Aristotle."' Yet McDougall's lectures drew large crowds, and were the only lectures at which women predominated over men. By now the McDougall persona had emerged – the leonine head, classical profile, longish hair and noble expression. Burt thought that 'his fine face and bearing and almost Olympian manner seemed immediately to set him apart as a Great Man – an effect which he secretly deplored'.[15]

Student and supervisor 'used to meet, almost in secrecy, like a pair of conspirators, in a large room in a deserted building at an hour when all healthy-minded undergraduates were "on the river"', Burt later recalled. 'His air of lonely magnificence did little to dispel the sense of solitude.' To carry out experiments, Burt sometimes bicycled to McDougall's house at Boar's Hill. 'There he became unexpectedly human and homely.' At this time McDougall had become interested in hypnosis and the subconscious mind and was carrying out experiments designed to demonstrate that in the hypnotic state memory and other intellectual powers might be apparently improved. He would then investigate whether the apparent improvement was due to the concentrated attention which such states seem to entail or merely to the effects of suggestion. McDougall was unable to hypnotise Burt, who later remembered 'roaring with laughter at this dignified don waving his fingertips before my sleepy eyes, like Svengali on the stage'; but he had more success with another of his students, May Smith. He did not publish his results, however, fearful that, given the popular conception of hypnosis as something occult, to do so might bring disrepute on his work. (As it was, when word emerged in Oxford that he had been working on hypnosis he nearly lost his job.) But they led to a general theory of sleep, suggestion and hypnosis which had a profound influence on British psychology.[16]

McDougall spent much of his time at home, delighting in observing the progress of his five children. It may have been the influence of the domestic setting which prompted him to change direction. It may also have been his urge to reach a broader public, having 'begun to realise that I was throwing my seed on stony ground, that my work along the lines I was pursuing could not find a public'. But it was probably also the climate of the times.

The British psychologist Philip Vernon once described the 1900s as 'the most exciting decade in the history of psychology since the death of Aristotle', referring to the period when Alfred Binet developed intelligence testing in France, Sigmund Freud and Carl Gustav Jung established psychoanalysis, and Ivan Pavlov was experimenting on his dogs. But there were also other forces at work which are more difficult to reconstruct today.

George Bernard Shaw – born in 1856, three years before the publication of Darwin's *On the Origin of Species* – described his age group as 'a generation which . . . began life by hoping for more from Science than perhaps any generation ever hoped before, and, possibly, will ever hope again'; but by 1909 he acknowledged that 'we are passing through a phase of disillusion'. Queen Victoria's reign had been a triumphant Age of Science: advances in geology, biology and physics had not only affected the way men thought about their physical environment but had altered men's ideas of their relationship to their environment – they had created a new cosmology and a new conception of the nature of change. By the 1900s, however, a reaction had set in. 'The reduction of spiritual facts to physiological phenomena', writes the cultural historian Samuel Hynes, 'may have made metaphysics obsolete, but it had not destroyed men's metaphysical itch, and much of what one might generally call Edwardian science is concerned with the problem of restoring metaphysics to the human world.'

Hynes argues that 'one field of scientific investigation in particular seemed to offer an escape from Darwinism – the field of mental events'. Because academic psychology had developed as a biological science, concerned with the physiology of the nervous system and subject to the laws of evolution, it had failed to address many of the issues which the broader public regarded as 'psychology'. It had therefore found its role usurped by less disciplined and less professional scientific writers, such as the sexologist Havelock Ellis, the essayist Edward Carpenter and the novelist Grant Allen.[17]

William McDougall's psychological writings of the 1900s represented an attempt to bridge this gulf. Misleadingly titled, *An Introduction to Social Psychology* was an ambitious attempt to provide a comprehensive scheme of individual psychology which could be applied across the social sciences; it was also an attempt to usurp philosophy's traditional role as the senior discipline. McDougall's intention was to replace the old static, descriptive and analytic psychology with a 'dynamic, functional, voluntaristic view of mind'; and his instrument for doing so was the theory of instincts. The idea came to him while lecturing one day in 1906:

> I found myself making the sweeping assertion that the energy displayed in every human activity might in principle be traced back to some inborn disposition or instinct. When I returned home I reflected that this was a very sweeping generalisation, one not to be found in any of the books; and that, if it was true, it was very important. I set to work to apply the principle in detail, becoming more and more convinced both of its truth and of its importance.

According to McDougall there were eleven instincts. Seven of them were linked to the primary emotions: the instincts of flight (fear),

repulsion (disgust), curiosity (wonder), pugnacity (anger), self-abase-
ment (subjection), self-assertion (elation) and the parental instinct
(tenderness). The remaining instincts played lesser roles in the
emotions: the sexual instinct, the gregarious instinct, and the instincts
of acquisition and construction.

Instincts were not quite the novelty McDougall claimed: both
Friedrich Nietzsche and William James had developed theories along
these lines. But his was by far the most systematic and wide-ranging
attempt to date to base social psychology upon a theory of instincts.
McDougall made sweeping assertions for the book and for his discipline. 'Psychologists', he wrote, 'must cease to be content with the
sterile and narrow conception of their science as the science of
consciousness, and must boldly assert its claim to be the positive
science of mind in all its aspects and modes of functioning, or, as I
would prefer to say, the positive science of conduct or behaviour.' His
belief that he had formulated a new theory of action, conduct and
character and provided a psychological foundation for ethics and the
social sciences met with fairly general acceptance. 'Culminating the
British Darwinian tradition of instinct-based theorising and integrating
it with a theory of emotion, its immediate impact is no mystery', the
historian Graham Richards has written. 'It must have struck contemporaries as a major step towards the kind of "total" theory that
Twentieth Century "scientific" psychology promised.' Over the next
two decades, An Introduction to Social Psychology would pass through
twenty-one editions, and it was still in print in the 1960s.[18]

The book's central argument was simple. 'Men are moved by a
variety of impulses whose nature has been determined through long
ages of evolutionary process without reference to the life of men in
civilised societies.' Moreover, 'the springs of all the complex activities
that make up the life of societies must be sought in the instincts and
in the other primary tendencies that are common to all men and are
deeply rooted in the remote ancestry of the race'. If, however, human
nature and society were generally determined by irrational instincts,
it was all the more important that the rational few should control the
mindless many for their own good. Like many Edwardian intellectuals,
McDougall feared the democratic masses. He was one of those excited
when the scientific polymath Sir Francis Galton called in 1904 for a
national debate on the issue of 'eugenics', which Galton defined as

'the study of the agencies under social control which may improve or impair the racial qualities of future generations either physically or mentally'. In the climate of national alarm after the Boer War, with worries about the physical degeneration of the race, preoccupation with 'national efficiency' and worries about the perceived failure of 'environmental social policies', Galton's words were eagerly taken up.

At the end of 1906, a satirical journalist noted that 'Mr Sidney Webb and the Fabian Society' had been 'directing their attention to our declining birth rate, and have come to the conclusion that, unless the profession of motherhood is municipalised, there is great danger of "national deterioration or, as an alternative, of this country falling to the Irish or the Jews" – and ultimately possibly even to the Chinese – presumably on account of the superior fertility of these races'. Webb's concerns were, he argued, misplaced; it wasn't so much that the birth rate was rising as that the death rate was falling.[19]

The 'race suicide scare' rumbled on throughout the Edwardian period, gathering new momentum when it emerged that it wasn't just that the British were being out-bred by foreigners; the wrong sort of Britons were having large families. 'The birth rate has decreased only in the upper classes, among the people with money and brains, the very people to produce the right sort of children, to tend them and educate them properly afterwards', wailed the popular writer Mrs Alec Tweedie; whereas 'the lower we go in the social scale the more prolific the people, and often the more undesirable the progeny'.

She concluded that 'the brain of the country is not keeping pace with the growth of weak-minded imbecility and vice. People do not seem to realise that we are rearing a race of degenerates at so rapid a rate that the future of Great Britain is imperilled.'[20]

McDougall soon joined the debate. In a paper read at the London School of Economics on 21 February 1906 he put forward 'a practicable eugenic suggestion'. Instead of taking 'negative' measures against the worst elements in society, 'the hereditarily criminal and degenerate' (that is, weeding out the least fit elements in the population), he proposed 'positive' steps to encourage the best elements in society, 'those of eminent civil worth', to have larger families. For example, senior civil servants (then earning about £800 a year) should be paid on a sliding scale, with bachelors getting only £500 and those with children more – he suggested up to £1,050 for a married man with six children.[21]

No one in his audience thought the scheme remotely practicable. Galton dismissed it out of hand. Other critics said that he had misunderstood how genetics worked; or they predicted that working-class birth rates would fall as affluence spread. The chairman of the meeting, Dr Frederick Mott, the pathologist at Claybury Asylum, thought that with better education the industrial working classes would start to 'furnish the brains of the country. They have already sent to this Parliament some very able men.'

There was, however, a further objection. McDougall had assumed that 'when the conditions are made clear to all, the really superior women' would abandon all thoughts of a career and 'once more find their highest duty and pleasure in producing, rearing and educating the largest number of children that their health and their means will allow'. Unfortunately, however, while his wife Annie might happily raise a brood, modern, intelligent women were reluctant to bear many children, particularly when told to do so by men. 'There is a real revolt amongst women against bearing as many children as their mothers and grandmothers bore', asserted the feminist Irene M. Ashby MacFadyen in the periodical the *Nineteenth Century*. In the United States, 'even the lowest classes of immigrants fresh from the slums and hovels of Eastern Europe begin to put a limit to their child-bearing as one of the first results of their new environment' and she had been surprised to find that 'the same feeling and the contingent precautions are widespread':

> The woman of today suffers more than her ancestors both in the anticipation and the hour of childbirth – that is the price paid in nerves and physique for her more complete and sympathetic share in the work, the thoughts, and the fortunes of her husband and children, and for the training that makes it possible.

African women might cheerfully bear child after child, 'but the white woman agonises over her child. The more intelligent she is, the more she knows of child life and hygiene, the more ambitious of the best, the intenser grow her cares and anxieties.'[22]

In the end, McDougall's proposal went nowhere. But he remained broadly in sympathy with the aims of the eugenics movement, arguing that psychology had as important a role to play as biology in implementing eugenicist ideas. The development of mental testing in the 1900s now made it possible to assess intelligence quantitatively – a task urgently needed because the 'mental qualities of the race' seemed to be most threatened by 'the conditions of high civilisation'. Children should be tested at the age of eleven or twelve, he argued, before the full advantages of social position had emerged, to identify those likeliest to benefit from better education. Psychology should also be used 'in the service of eugenics in the direct attack upon the problem of mental heredity'.* And in the approaching era of 'universal miscegenation', mankind, 'conscious of itself as a whole, [should] take intelligent thought for its own future and attempt to regulate in some manner and degree this process of racial mixture'.[23]

If eugenics was McDougall's 'No. 1 scientific hobby', psychic research ran it a close second. By the 1900s the Society for Psychical Research, founded in 1882 by a group of Cambridge dons, had become a powerful force in English intellectual life. Early supporters included Lord Tennyson and Mr Gladstone, respected scientists such as Sir Oliver Lodge played an important part in its activities, and the society was scrupulous in its testing of evidence and ruthless in exposing frauds. For all that, psychic research owed much of its success to the fact that it filled the gap left by the decline of religion.[24]

The one community which tended to be most hostile to psychic research was the small cadre of academic psychologists, who saw their efforts to claim scientific respectability for their discipline being constantly undermined by spiritualist groups who also used the terms 'psychology' and 'psychological'. Surveying the progress of British science in 1906, the biologist Ray Lankester was keen to distinguish between 'enthusiasts who have been eagerly collecting ghost stories and records of human illusion and fancy' on the one hand, and 'the serious experimental investigation of the human mind, and its

---

* Caleb Saleeby, one of the moving spirits in the launching of the Eugenics Education Society in 1907, claimed to have inspired McDougall's *Introduction to Social Psychology* by urging him to clarify the tangled question of inheritance by providing 'an outline analysis of what is really inherent, and therefore alone transmissible, in the human mind'. C. W. Saleeby, *Parenthood and Race-Culture* (1909), p. 117.

forerunner, the animal mind, which [had] been quietly but steadily proceeding in truly scientific channels', on the other.[25]

Why then did William McDougall take psychic research seriously? Simple perversity was certainly one explanation – the desire to distinguish himself from, and to annoy, his professional colleagues. Then again, his hero, William James, had become heavily involved in psychic research after becoming impressed by Leonora Piper, a spirit medium in Boston, in 1885. There is, too, evidence that McDougall's wife Annie was interested in this side of his work. Probably, also, McDougall's craving for greater public recognition led him to reach out to a wider readership through psychical research. But the overwhelming reason seems to be that by 1908 he had become convinced that the 'mechanistic psychology', as he now routinely called it – that is, a purely scientific approach to the mind – failed to explain and illuminate many aspects of human behaviour and conduct. Psychic research, by contrast, might provide a window to understanding those unconscious areas of the mind revealed by dreams and hypnosis. A true scientist should, therefore, have an open mind about these matters.[26]

McDougall set out this view at some length in *Body and Mind* (1912), which he later described as the most extreme example of his 'uncompromising arrogance'. Subtitled *A History and Defence of Animism*, the book combined an extended critique of mechanistic modes of thought in the life sciences with a long review of animism, the view that the human body is given life by something non-material – what had traditionally been called a soul. McDougall was well aware that 'souls were out of fashion', as William James had put it, but he 'had a predilection for unfashionable doctrines. And, seeing that so many scientists seemed to be finding satisfaction in shocking the bourgeois, I would shock them by putting up a defence of an exploded superstition.'[27]

McDougall believed that occurrences such as telepathy and automatic writing could not be reconciled with 'the mechanistic scheme of things'. Hypnotic phenomena, too, he contended, largely remained 'refractory to explanation by mechanical hypotheses', and went far to uphold the belief that ordinary organic processes 'are in some sense controlled by mind, or by a teleological principle of which our conscious intelligence is but one mode of manifestation among others'. Yet the conclusion of *Body and Mind* was cautious. McDougall accepted that most of the evidence that 'human personality may and does survive in some sense and degree the death of the body' was inconclusive. On the other hand, he felt that psychical research had established the existence of phenomena that were incompatible with mechanistic psychology. It had also shown very clearly how certain states of mind could produce corresponding physical symptoms.[28]

What drove McDougall's phenomenal output of words? What was the source of his enormous literary energy? If we were to engage in amateur psychology, we can see several contradictory instincts at work: a childish instinct to outrage, annoy and subvert his academic peers; a craving for fame and public position; a need to make and preserve his home; an instinct of survival; a feeling that the position of his race (and his class) was under threat; a wish to reach a wide readership and a need to be academically respected. It was gratifying that his books sold in huge quantities, yet there remained a frustration, a sense that, having aspired to the status of a great man and major public figure – and spent decades mastering philosophy, physiology and psychology – he was not achieving his purpose. In his autobiographical essay, McDougall recalled that the publication of *Body and Mind*, like that of *An Introduction to Social Psychology*, 'was like dropping a stone into a bottomless pit':

> I waited to catch some reverberation; but in vain. Each book received, I think, one favourable mention in the press; and that was all. I never could discover that anyone in Oxford had read either of them. And my colleagues, with one or two exceptions, seemed to be shaking their heads very gravely. About this time I began to find it difficult to believe in the value of my work, a difficulty that has grown steadily greater. I was much tempted to turn to medical practice before it should be too late.[29]

However, in 1912 things improved. McDougall was elected a fellow of the Royal Society, largely on the strength of his experiments on vision, and finally secured a bridgehead in Oxford by becoming a fellow of Corpus Christi, a small and scholarly college mainly notable as an outpost of pragmatism, the movement in philosophy espoused by William James. Pragmatists rejected the idea that the function of thought is to describe, represent, or mirror reality or 'truth'. Instead, they contended that most philosophical topics – such as the nature of knowledge, language, belief, and science – were all best viewed in terms of their practical uses and successes rather than in terms of representative accuracy. At Oxford, however, pragmatism was not the most pragmatic option: it was generally acknowledged that because F. C. S. Schiller, the Corpus tutor, was the leading 'pragmatist' when 'Idealism' provided the dominant philosophical doctrine in the university, for a generation nobody from Corpus got a first in Greats. 'In the 1902 Greats examination three Corpus finalists happened to be sitting in a row. As he passed the third of these Corpus desks during a philosophy paper, an examiner who was invigilating, noting the college's name on the desk labels, groaned audibly, "Oh God, nine more hours of pragmatism." The Corpus men concerned thought this a great joke.'[30]

For McDougall, however, the road was now not quite so lonely. Schiller, an entertaining, black-bearded man, famous for his lavish Sunday breakfasts, shared his interest in eugenics and psychic research. 'My position was thus greatly strengthened; and I felt a certain obligation to persevere in the Paths of pure science, however little I might effect', McDougall recalled. In 1912 he produced *Psychology, the Study of Behaviour*, a small volume for the Home University Library, in which he forcefully restated his view that mechanistic models of the brain could not explain human behaviour and, in particular, what he now regarded as the dominating character of human life – purposive activity. The book would go on to sell 100,000 copies.

His thoughts now turned to producing the volume on group psychology, for which his *Introduction to Social Psychology* had been only the overture.

'It was Germans, Germans all the way'

Early in 1900 Charles Myers returned from the Torres Straits expedition to take up a position as house physician at St Bartholomew's Hospital in London. Like Rivers, he found the demands of that job, with its long hours and heavy responsibilities, very onerous; the following year he was forced to resign from the hospital and went to Egypt to convalesce – where, as we have seen, he met up with some of the other expedition members. He decided to give up medicine, and on his return to England, went to Cambridge to help Rivers with classes in experimental psychology.[1]

Over the next decade, as Rivers became more involved in anthropology, Myers gradually took over psychology at Cambridge. In 1904 he was appointed university demonstrator in the subject; that same year he also married Charles Seligman's cousin Edith. He combined his post with a professorship in experimental psychology at King's College London, until the strain of working in two places caused another breakdown in 1909. Thereafter, he concentrated on Cambridge, where he was promoted to lecturer.

Myers worked hard to build up the discipline. Although in the 1900s there were mighty tomes by William James and the British psychologist, G. F. Stout, there was no decent textbook in English; the surveys of experimental psychology were all in German. Myers read and digested all this literature and in 1909 produced a solid, lucidly written *Textbook of Experimental Psychology*, which at long last gave English students something to work from. (Typically, it steered a middle path between the views of Rivers and McDougall on the contested subject of colour vision: 'The evidence points to the existence of both a three- and four-colour basis of colour vision', Myers wrote. 'Some day, perhaps, we may be able to show how these two systems have been combined, one perhaps superadded at a later date in the course of evolution.') Two years later he produced a shorter version.[2]

Myers also managed to create a decent psychological laboratory, though it took time and money. This long-running saga went back to 1893, when Rivers had begun teaching in Cambridge in a room in the old physiology department, from where he migrated on returning from the Torres Straits to rooms in a building on St Tibbs Row. Then, in 1903, the university press offered the psychologists the use of a small cottage it owned in Mill Lane, not far from the banks of the Cam. The atmosphere was 'damp, dark and ill-ventilated' and the joke later went that it was 'a wonder that Behaviourism did not first grow here, instead of later and elsewhere' because 'the river was nearby, rats abounded and anybody could observe startle reflexes firing off in all directions'. In 1908, Myers launched an appeal for money to build proper premises and persuaded the physiology department to donate to the psychologists a wing of the magnificent new building they were planning to put up with money from the Drapers Company. He then raised most of the money for it himself – either from his relatives or out of his own pocket (his father had recently died). On 18 July 1911, his mother laid the foundation stone for the new building at a ceremony watched by numerous Cambridge worthies.[3]

Even then Myers remained wary. The following year he declined to host a meeting of the International Psychological Congress in Cambridge because the gathering would not be confined to academic psychologists; anybody could join it and read a paper, or at any rate join in the discussions. He was clearly worried that members of the Society for Psychical Research would attend. 'Already in England there is a general belief that anyone has the right to give his opinion on psychological matters', Myers explained to the Harvard psychologist Hugo Münsterberg. 'I believe that if the Congress meets in Cambridge, it will not only help to foster this belief but will do great harm to the progress of psychology in the University . . . so delicate is its position.'[4]

Although his time was much taken up by administration, Myers continued his work on music. He wrote up the Torres Straits material and, without himself venturing into the field again, began to interpret musical recordings brought back by others. He became the 'go-to man' at Cambridge, advising other anthropologists on how to make recordings and then analysing their findings when they returned. He

1. Scientists and their servants: group portrait on Murray Island, May 1898. (*Back row*) William McDougall, Charles Myers, Charles Seligman; (*second row*) William Rivers, Sidney Ray; (*first row*) Alfred Haddon, Charlie Ontong, Anthony Wilkin; (*seated*) Jimmy Rice, Debi Wali.

2. Rivers testing eyesight, Mabuiag, 1898.

3. 'I had a great morning of song': Charles Myers recording the sacred song of the Malu ceremonies, Murray Island, 1898.

4. Charles Seligman recording children in Hula, Papua New Guinea, 1898.

5. A piratical group: Alfred Haddon (*seated*) with (*left to right*) Rivers, Seligman, Ray and Wilkin on Mabuiag, autumn 1898. By this time, Myers and McDougall were en route to Sarawak.

6. The Resident and his people: Charles Hose with a group of tribesmen, Sarawak, *c*.1900.

7. William Rivers (*right*) and Henry Head conducting their experiments on the nervous system in Rivers's rooms, St John's College, Cambridge, *c*.1903.

8. William McDougall (*back row, second left*) as an undergraduate at Cambridge, May 1893.

9. Anne Amelia Hickmore, the Brighton chimney sweep's daughter McDougall married in a hurry. A portrait taken in Oxford around 1910.

10. McDougall the family man: on holiday in England in the summer of 1931 with his wife and (*clockwise*) daughter Lesley, and sons Duncan (killed in an aeroplane, 1932), Kenneth (killed in France, 1945) and Angus.

11. William Brown, William Rivers and Grafton Elliot Smith at Maghull War Hospital, near Liverpool, 1915.

12. Charles Myers; a studio portrait taken in Folkestone, 1916.

13. The first stage of medical evacuation: a dressing station at the Western Front, 1916.

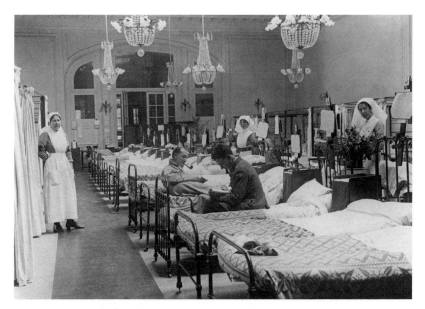

14. The Duchess of Westminster's Hospital, in the Casino at Le Touquet, where Charles Myers first identified 'shell shock'.

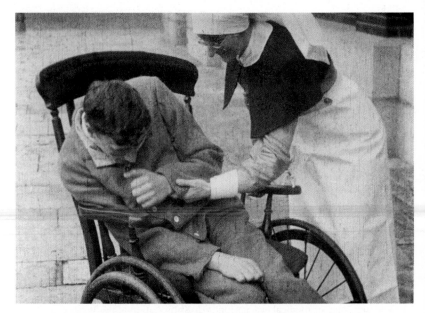

15. William McDougall's most famous patient: Private Percy Meek regressed to being a child after unsuccessful treatment.

16. Private Meek in 1918, after his spontaneous recovery, teaching basket weaving at Seale Hayne Hospital.

collaborated especially closely with his friend Charles Seligman, who was now emerging as one of the great field anthropologists of the day, especially after his marriage in 1905 to Brenda Salaman, a feisty young woman from a prominent English Sephardic family. In their joint studies of the Vedda tribe in Sri Lanka and of the Nilotic peoples of the Sudan, Charles Seligman carried on Haddon's tradition of assiduously gathering information, while his wife was able to observe rites and ceremonies among women which no man would have been permitted to witness.

In 1913 Myers published an article on 'the beginnings of music', which compared the Malu dances he had recorded on Murray Island with the music of the Vedda in Sri Lanka and the tribesmen in Sarawak. Painstakingly he identified the development of rhythm and tone, finding the Malu and Vedda music 'primitive', whereas on Sarawak external influences – Mongolian, Malayan and possibly Indian – had produced much greater sophistication. Characteristically, however, he declined to speculate on the origins of music, arguing that the rival hypotheses, 'which variously ascribe the beginnings of music to speech or to the imitation of the sounds of nature, or which lay stress on the importance of sex or of rhythm', simply reflected the individual preferences of modern writers.[5]

The comparative study of music – what we now call ethnomusicology – was then a completely new field, and Myers was a pioneer in it, combining deep understanding of the music itself with real knowledge of the societies which produced it. 'The musical work was so good, and at the time so original,' writes his biographer, 'that it is almost certain that if he could have devoted himself with a single mind to this kind of study, Myers would have achieved very great distinction indeed in the understanding of human musical performance and appreciation.' Yet, he adds, 'his contributions remain more of the nature of fugitive pieces, full of interest, full of promise, but never fully developed'.[6]

In February 1912, Myers gave a talk on 'Primitive Music' in Cambridge in which he sang some of the pieces he had collected and spoke about cultural differences in musical perception, such as the predominance in some aboriginal peoples of absolute pitch and the ability to perceive 'many successively different intervals of time as a coordinated whole – a picture'. A young Austrian student in the audience was so excited that he came and worked with Myers, then still in the dismal cottage in Mill Lane, devising experiments to ascertain the extent and importance of rhythm in music. They seem to have got on well together, though Ludwig Wittgenstein's youthful arrogance sometimes got the better of him. In June 1912 he confessed to Bertrand Russell that he had 'had a discussion with Myers about the relations between Logic and Philosophy': 'I was very candid and I am sure he thinks that I am the most arrogant devil who ever lived. Poor Mrs Myers who was also present got – I think – quite wild about me. I think he was a bit less confused after the discussion than before.' In fact Myers was quite used to being patronised by philosophers: in his early Cambridge days he had given a paper on 'vitalism' to the Moral Sciences Club and then been torn apart by the philosopher G. E. Moore; as a result he had completely abandoned the philosophic approach and never published anything in that area. He forgave the young Austrian – even if his wife did not – for when the psychological laboratory was formally opened in 1913, Wittgenstein exhibited an apparatus for the psychological investigation of rhythm.[7]

For all his modesty, Myers could convert people to his discipline. Tom Pear, a student of physics at King's College London, 'became convinced, when attending Myers' lectures, of the crucial psychological issue underlying all science, that scientific observation implies an *observer*'. He later recalled that 'a great change suddenly happened to my "world-view" [and] this caused a swing over from a belief that the essence of everything was physical and chemical, to one that physical scientists, in their attempts to penetrate the reality of the known, were deliberately ignoring the *knower*. I fell out of love with physics and chemistry.' Pear found Myers an attractive personality: 'unusually many-sided: doctor, anthropologist, musician, Alpinist, traveller. Few psychologists had been to so many places.' He also valued Myers' humanity: 'Though a fine experimentalist, he never believed that the most important things in life could be experimentally investigated or measured. There would never have been any need to remind him that all human psychology is social psychology or that society is made up of individual persons.'[8]

Myers' other protégé, a young man from Gloucestershire called Fred Bartlett, later recalled the 'growing excitement' with which he had read the *Textbook of Experimental Psychology* in 1909, the year of its appearance. 'I read and reread its pages, admired the unornamented but lucid, graphic and economical style of its writing and the sincerity and authority which the book carried with it.' Bartlett found Myers 'rather variable as a lecturer, deliberate, often even hesitating, and normally relied upon carefully prepared but sparse notes . . . a lecturer for those who preferred matter to fluency. He was at his best in the conversational manner and the small class. In front of a large audience he often appeared ill at ease and vacant.'[9]

At the end of the decade, Myers hired as his lab assistant McDougall's pupil Cyril Burt, even though, coming from Oxford, Burt advocated some controversial theories about the interaction between body and mind, the importance of unconscious and purposive mental forces such as the instincts, the value of mental eugenics, even the applicability of relational analysis to the mind. Burt quickly found that, beneath Myers' air of critical reserve, he possessed 'not only a most kindly tolerance for the heresies of the young, but a surprising sympathy with the very tendencies that he so often questioned, and a secret eagerness to press on towards practical applications'.[10]

Nevertheless, the ultimate direction of his sympathies was clearly shown in the *Textbook*, where two of the seven chapters were devoted to mental tests and the rest given over to the standard German experimental fare. Although Myers talked to students of a 'sociological psychology in the future', he insisted that the foundations had to be laid first: 'you can go on later with your upper storeys – educational, industrial and then medical'. The right order was: 'The primitive mind first or the child mind, if you like; then the industrial mind; and the abnormal last of all: that seems to me the natural order, since each in a sense implies the last.'[11]

A young woman graduate student who turned up in Cambridge in 1909 to do psychology was surprised to find that the course was run by 'a young man of thirty-six'.

He was maybe a bit above average height, slightly stooping but lithe and active (he was a first-rate dancer and was fond of it), very friendly and approachable. He had a noticeably large head, fair hair and eyes grey with a suggestion of blue, the hair already thinning a bit. His face in repose was somewhat heavy, contemplative, but he had a most lovely and winning smile which, at this time was frequent and altogether captivating.[12]

What was the curriculum at Cambridge at this time? According to Bartlett, the course which Cyril Burt taught, under Myers' supervision, was traditional brass-plate psychology. It covered the classic experiments and used the apparatus devised by all the great names of the past: Hermann von Helmholtz and Ewald Hering on colour theory; Wundt; Blix, Goldschieder and Frey on sensory spots in the skin; G. E. Müller on vision; Kraepelin's ergograph for testing fatigue. There were, though, 'one or two little refreshing excursions into more recent work such as the Würzburg school' (a heretical group which in the 1900s tried to make experimental psychology reflect more the subjective workings of the mind) and 'a little, a very little, of the new psychology of mental tests, and of Jung's form of word associations'. In short, 'It was Germans, Germans all the way, and if we were going to stick to psychology then to Germany sooner or later we must all surely go.' Bartlett found it rather tedious:

> Privately we grumbled and groused like anything. We vowed we would lift no more weights, learn no more nonsense syllables, strike no more tuning forks, cross out no more e's. Colour wheels were more fun. We threw ourselves at the dynamometer and the ergograph in the hope of beating one another or smashing the apparatus . . . When we left the laboratory we promptly dropped it all. What had it to do with our daily life? It was indeed a laboratory game, boring in parts, engrossing in others, but just a laboratory game. And it was not very easy to link up the experiments with the general teaching.[13]

Charles Myers and his growing family lived the comfortable life of the Edwardian don in a large house in the village of Great Shelford, just outside Cambridge on the Gog and Magog Hills, with a huge garden. They entertained frequently and Myers played the violin. From time to time, his pupils were invited to lunch, tea, or dinner. 'He and Mrs Myers were the most wonderful hosts', Bartlett recalled. 'We loved it. These four or five years while Europe was blundering on towards the first of the two great wars were, I think, Myers' happiest time in Cambridge, perhaps in his life. He was enormously busy, full of plans for the new psychological laboratory, he had some first-class students, his reputation was growing rapidly overseas, he foresaw a great future for psychology and he had a lovely home life.'[14]

13

'A great part of the truth'

William Rivers, William McDougall and Charles Myers all played a prominent part in the activities of the British Psychological Society, which, as we have seen, McDougall and Rivers had helped to start in 1901. Myers, abroad when the society was founded, became secretary in 1904; in 1911 he also became co-editor of the *British Journal of Psychology*, before serving as sole editor from 1914.

The society grew slowly over the next decade, but was careful to restrict membership to professionals, physiologists, neurologists and psychiatrists as well as academic psychologists. It quickly fell into the routine of holding monthly meetings, usually on a Saturday afternoon in London, where two or more papers were spaced by a tea interval and the meeting was followed by a dinner at Pagani's restaurant, in Great Portland Street. Myers' protégé Tom Pear, by then living in Manchester, was a regular attender:

If one lived in the north, to attend an afternoon meeting of the Society in London it was necessary to rise early. So it might be daunting to find at King's College at 2.30 p.m. no signs of life except an aloof cat and a bonhomous hall porter. He would assure you that the meeting was certainly 'on', but 'the gentlemen are never very punctual, sir; *you* understand'. Sure enough, about 2.35 p.m. a small chorus of dons could be heard advancing up the corridor . . . and up would stroll the in-group, to be augmented by stragglers, until at about 2.45 p.m. there might be thirty in the meeting-room. Not many, but representing philosophy, education, neurology, psychiatry and anthropology, and perhaps a third of the audience were already known in the English- and German-speaking worlds of knowledge . . . And – for in those days the Society was a very British one – we confidently expected, and got TEA.[1]

Rivers gave papers on 'The influence of alcohol on muscular and mental efficiency' and 'Visual acuity in different races of men' and (with Head) on 'Some points of psychological interest suggested by a case of experimental nerve division'. Myers spoke on 'Theories of consonance and dissonance', 'The rhythmical sense of primitive people' and 'Observations on contrast with smoothly graded disks'. McDougall offered 'The bearing of modern experimental work on the problem of the unity of the mind', in 1904; 'The fundamental forms of mental interaction' in 1906; and 'Some difficulties connected with the current conceptions of instinct' in 1909.

The dinner afterwards was organised by Alexander Shand, a lawyer and amateur psychologist, whose conversation and taste in wine were fondly remembered years later, and who arrived always in a spick and span navy blue suit. Although the headmistress of the North London Collegiate School had been one of the founder members of the society, it was not the custom in its early years for women to attend the dinner. They were never excluded from these dinner gatherings but 'it seemed to them wiser not to attend', Dr Beatrice Edgell recalled in 1947. 'If today this point of view is difficult to understand one must see in it a measure of the change in social relations which time has brought about. One thing is clear to this writer: had the women members attended those early dinners would have been more formal and the discussions less spontaneous.'[2]

Just occasionally, in the minutes of the society's meetings, a hint of discord can be detected. In 1908 the rivalry between Myers and McDougall surfaced in a comic way. McDougall gave a paper on the colour sense of two of his children – he had been interested to discover at what age they began to be capable of appreciating colour and the order in which the primary qualities of colour sensation were developed, testing them with coloured flowers and balls. Perhaps unsurprisingly, his observations faithfully illustrated his own theories of colour vision. This prompted Charles Myers to produce a paper based on observing the development of the colour sense in *his* elder child, using coloured bricks; but he then went on to question, perhaps typically, whether the development of colour sense could be tested at all.[3]

The following year there was a more serious disagreement. The psychologists became involved in the national debate over the state of Britain's population which had been provoked by the military reverses of the Boer War and by the revelation (in the recruitment process) that British men seemed to be becoming smaller. If their bodies were shrinking, what was happening to their minds, many wondered. The British Association for the Advancement of Science created a committee to assess the possibility of measuring the entire country, physically and psychologically, and by 1908 Myers and McDougall both sat on a subcommittee which strove to find a suitable method of 'measuring mental characters'. As the decade had progressed, the arrival of techniques for testing intelligence, especially those developed by the French psychologist Alfred Binet, made this a real possibility. McDougall was an enthusiastic supporter of this work and, with the help of the Board of Education, sent his assistants Cyril Burt, J. C. Flugel and Horace English to investigate the intelligence of children from different social backgrounds in Oxford, Liverpool and elsewhere. At a time when the eugenics movement was campaigning for the sterilisation of the feeble-minded, this was a highly contentious area.[4]

Myers, characteristically, was more cautious. In 1911, he published an article questioning the value of mental testing. His primary objection was to the idea, which seemed to be fast gaining ground, that masses of psychological data could be collected by an army of untrained observers. Experience had shown that even in physical anthropology there were marked discrepancies between the results obtained by different observers, and by the same observer at different times, so the idea that you could standardise mental tests and have them used by untrained personnel should be resisted. But Myers' objections went beyond methodology:

I shall be asked, Are not these striking and new results which have lately been obtained by statistical methods, showing the absence of correlation between the state of nutrition of school children and their mental capacity, or between alcoholism in the parent and defective health in the offspring? And I reply that, in my opinion, these results have no real value. They have been obtained by applying scientific methods to the solution of a problem of such complexity that the solution appears in the form of a meaningless blur.[5]

The gentle self-satisfied world of the British Psychological Society was about to be shattered by thunderbolts. The first of these, the arrival of psychoanalysis, can be described in two ways. The simpler, yet more dramatic, version is that in 1910 the *American Journal of Psychiatry* devoted an entire issue to the lectures which Sigmund Freud and his associates Carl Gustav Jung and Sandor Ferenczi had recently given at Clark University in Massachusetts. Suddenly, American and British doctors and academics were asked to accept that the analysis of dreams provided the royal road into the unconscious, that the origins of all neuroses could be found in sexual conflicts, and that children had sexual lives. The older psychologists were revolted by Freud; the younger ones, fascinated and excited.[6]

The second version of the story is more complex. If we look at what the members of the British Psychological Society did *not* discuss at their meetings, we get an insight into the parochial, insular intellectual world they inhabited. Thus, for example, the minutes make no mention of the work of the German psychiatrist Emil Kraepelin in categorising psychological illnesses, which is now regarded as one of the foundations of modern psychiatry. Neither was anything said at the meetings about Eugen Bleuler, the Swiss psychiatrist who coined the term 'schizophrenia', to replace dementia praecox. These omissions confirm the point Rivers made when he worked as a psychiatrist in the 1890s: that in Britain fruitful intellectual contact between experimental psychology and the mental health system, which was such a feature of French and German medical life (as Rivers had seen in Jena and Heidelberg), simply did not happen. This mattered because the end of the nineteenth century proved to be a period when insights into the workings of the normal mind were gathered by studying the abnormal one. One important strand, which stretched back to J. M. Charcot's work at the Salpêtrière Hospital in Paris, and was continued by his pupils and successors Joseph Babinski and Jules Dejerine, involved the study of hysteria; another was the use of hypnosis by a group of doctors in the French provincial town of Nancy.

These influences came together in the work of the most significant pioneer in the field before the arrival of Freud – the French psychiatrist, Pierre Janet. Philosophically trained and a pupil of Charcot, Janet worked extensively with seemingly ordinary people who had troubles in their everyday lives. 'Hysterics', he wrote, 'are patients who are easily managed, who talk willingly, who are not dangerous, on whom one can experiment without any great fear, and who, lastly, like to observe and readily lend themselves to observations.' He discovered, mainly by the use of hypnosis, that his patients' problems were usually traceable back to some terrible event in the past. This led him to the concept of dissociation – the splitting of the mind. Janet argued that consciousness does not necessarily consist of a single and homogeneous stream, but that it is sometimes split into a number of more or less independent currents, and that the dissociation thereby produced explains a large number of phenomena, not only in hysteria and other disorders, but in the normal processes of our daily life. Janet called his method analytical psychology.[7]

In his Clark lectures, Freud acknowledged his debt to Janet – 'we took the splitting of the mind and dissociation of the personality as the centre of our position' – but then emphasised how far he had moved things on by providing an explanation for *why* the mind dissociated – or, in Freudian terms, 'repressed' – memories. His argument was that the mind repressed the memory of certain events, if they involved 'a wish that was incompatible with the patient's ego'.[8]

Some elements in British intellectual life were familiar with Freud's work long before 1910: most notably the Society for Psychical Research had made contact with him (and with Janet) before 1900. And although hysteria was not 'cultivated' by British doctors as it was by French, there were neurologists who made it their business to keep up to date with French writing on the subject. By the same token, it would be a mistake to think that psychoanalysis swept all before it in Europe; in fact, it was rejected by mainstream psychiatric opinion in Germany and Austria.[9]

Such, then, was the situation when Freud's lectures appeared in the *American Journal of Psychology* in 1910. A generation gulf quickly emerged. Tom Pear recalled that 'There was daily talk of [Freud] among the younger English psychologists, but I doubt if the new ideas disturbed the oldsters much.' William Brown, McDougall's first pupil, wrote in the *Strand Magazine* in January 1913 that 'The explanation of dreams given by Professor Sigmund Freud, of Vienna, is exceptionally original, as well as being highly ingenious and interesting. Even if it is not the entire truth about dreams, there is little doubt that it contains a great part of the truth.' The same year, Pear told the British Association that while Freud's theory of dreams did not hold good for all dreams, 'the "dream-work" and the "censorship" exist and play an important part in mental life. Freud's striking demonstration of them is a valuable contribution to psychology.'[10]

William McDougall found himself in an odd position – involuntarily cast as John the Baptist to Freud's Messiah. In some ways he had prepared the way for the Viennese, championing a dynamic psychology, exploring the subconscious mind, and arguing that human behaviour was ruled by instincts; and, having himself abandoned scientific caution and embarked on speculation about the workings of the mind, he could hardly object when others speculated too – albeit more boldly. At first he seemed uncertain how to respond. In *Body and Mind* he acknowledged that 'the success of Freud's therapeutic methods in his own hands and in those of a numerous and rapidly increasing band of disciples proves that there is a large basis of truth in his doctrines' and in *Psychology* (completed in February 1912) he paid tribute to both Janet and Freud, whose work had 'enriched the field'. But by 1914 he had become much more critical. A note on 'the sex instinct' in the latest edition of his *Social Psychology* accused Freud of confusing the sexual instinct with the sentiment of love, and of extending to normal people generalisations which were true only of a certain number of individuals of somewhat abnormal constitution, from among whom his patients had been drawn. He then rejected the Freudian doctrine of infantile sexuality, drawing on Darwin, studies of animal behaviour and the observation of his own children to argue that 'it is at about the age of eight years that the behaviour of children commonly begins to exhibit indications of their attraction towards and a new interest and feeling towards members of the opposite sex'. He was especially vehement about Freud's doctrine of erogenous zones: it was 'very obscure, seems incapable of being rendered clear and self-consistent, and betrays a conception of the nature of the sex instinct which is vague, chaotic, and elusive, uncontrolled by consideration of the facts and inconsistent with these facts'.[11]

He expanded on these arguments publicly at a meeting of the Royal Society of Medicine in March 1914. McDougall acknowledged the contribution made by German medical writers on sex, such as Albert Moll, Richard von Krafft-Ebing, Iwan Bloch and Freud himself, and accepted that mainstream psychology had neglected sexual factors – William James had hardly touched on sex – but he argued that there were two great errors in psychology: the error that instinctive action is merely compound reflex action; and, secondly, the error of psychological hedonism, the doctrine that pleasure and pain are the prime movers of all human and animal behaviour. Freud, said McDougall, was guilty of both errors. 'The Freudians conceive the innate basis of the sexual life altogether too narrowly' and 'hasten to attribute to the sexual instinct large numbers of mental and bodily activities which are rooted in other instincts than the sexual, or are highly intellectualised processes determined not by one instinct, but rather by highly complex sentiments, in which perhaps the sex instinct has no part'. He concluded that while 'Freud has done and is doing a great work for the furtherance of psychology', the Freudian doctrines required 'very considerable modifications'.[12]

The younger doctors in the audience were not convinced. William Brown came up with what was fast becoming one of the key rejoinders to Freud's critics – that if you hadn't used psychoanalysis you were unable to judge it.

By this time, however, it had become clear that McDougall's reservations were shared by some of Freud's own lieutenants. In 1911 Alfred Adler had been expelled from the psychoanalytic movement because he 'minimised the sexual drive'. Much more seriously, the following year Jung, Freud's 'crown prince', doubly valued both as a Gentile and a well-respected psychiatrist, had begun to make public his own differences with the master. Jung, too, had doubts about the sexual origin of neurosis, did not agree that madness was caused only by psychological factors and wanted Freud to bring psychoanalytic theory into closer alignment with biology. Returning to Europe after lecturing in America in 1912, Jung bluntly told Freud that his own version of psychoanalysis had 'won over many people who until now had been put off the problem of sexuality in neurosis'. By the beginning of 1913 the two men had ceased to correspond and Jung had begun an offensive aimed at taking his gospel to the Anglo-Saxons. From his perspective, the support of a prominent British psychologist such as McDougall was important.[13]

In August 1913, Jung appeared at the International Medical Congress being held in London, debating with both Janet and Ernest Jones. He also began to develop an alternative to one of Freud's central ideas, that of the libido. Whereas, for Freud, the libido was closely related to sexuality, in Jung's new formulation the libido was something much wider, more like a life force, vital energy in general. The *British Medical Journal* hailed his new approach as a return to 'a saner view of life'. Ernest Jones began to be seriously worried that several prominent British Freudians would defect to Jung.[14]

William McDougall was in the audience when Jung addressed the Psycho-Medical Society in London on 24 July 1914. He heard Jung suggest that the word 'libido', which was confusing and associated with Freud, should now be replaced by the term 'hormé' and that 'psychoanalysis' be renamed 'prospective psychology'. In fact, Jung himself never pursued either of these phrases, though urged by McDougall to do so, and it was McDougall who later took up the term 'hormé' and labelled his system 'hormic' psychology.[15]

There was a second consequence of the talk. 'Jung unfortunately had a great success in his London lecture,' Ernest Jones reported to Freud, 'and McDougall was so impressed that he is going to be analysed by him.' The Freudians recognised that in winning the approval of Britain's most prominent psychologist their rival had secured an important scalp. McDougall had a long private meeting with Jung and seems to have responded warmly to his personality, though no record of the conversation survives and there are some grounds for doubting Jung's sincerity. He seems to have had a low opinion of British intellectual life. While in London he had also attended a symposium on 'Are Individual Minds contained in God or Not', organised by British psychologists and philosophers and, according to his biographer, 'found the proceedings intellectually shallow and disappointing and listened in astonishment to arguments that would not have been out of place in a thirteenth-century monastery'.[16]

Unaware that he was being used, McDougall arranged to visit Jung in Zurich in order to be analysed by him. His life seemed to be about to take an important new direction.

But in August 1914, the second thunderbolt arrived: the assassination of a Habsburg archduke escalated into a European conflict. Jung slowly made his way back to neutral Switzerland, through a Germany gripped by war hysteria. McDougall plunged into a new battle.

PART FIVE

THE WOUNDED MIND

14

'This orgy of neuroses and psychoses and gaits and paralyses'

For several years it was Charles Myers' custom to spend the month of July climbing in the Swiss Alps. Returning through Paris in early August 1914, he found the city in a state of 'great excitement and turmoil'. It was hard to get a cheque cashed; the waiter on the train to Calais would only accept payment in gold or silver. Myers reached Cambridge on 4 August 1914, just as war was being declared.[1]

He went back to work, sorting and interpreting a collection of recordings of Australian Aborigine music which had been bequeathed to him. But, as the country mobilised for war, the British Expeditionary Force landed in France, and the first battles were being fought, this task seemed increasingly irrelevant and purposeless. By the end of the month Myers could bear it no longer. He went to London and offered his services as a doctor to the War Office, the Order of St John of Jerusalem, and the Red Cross. But nobody would have him – he was now forty-one years old, had not served in the Territorials and had not practised medicine for over a decade. His only hope, he was told, was to make use of social connections. So, when he heard that the Duchess of Westminster was assembling in France a medical unit composed of doctors from St Bartholomew's Hospital, Myers took the ferry to Paris. It was now 17 October 1914, and the First Battle of Ypres had begun.[2]

The next day, amid the 'general sadness and drabness' of the French capital at war, Myers met up with his former colleagues and assessed the situation. The medical arrangements were still being sorted out: there were several British volunteer units in Paris and it was proving difficult to assign them. The Duchess of Westminster's hospital, lavishly equipped by public subscription, had been kicking its heels in the Hôtel de France et de Choiseul for sixteen days, while members of its staff drove off to towns near the front trying to find a suitable building to use. 'All reports say the French are terribly short-staffed; yet these British surgeons with temporary army rank cannot help!' Myers complained in his diary. 'I decide to play a waiting game to see if I can find a job with [them].'[3]

He did not have long to wait. The hospital's matron, an old ally at Bart's, put in a word on his behalf. Soon he met the Duchess of Westminster herself, 'a very clever, bright and unaffected little woman and not at all bad-looking', who offered to do what she could for him. On 20 October, after being definitely promised work in her hospital, he took the bold step of ordering a uniform from the Paris branch of Burberry's; five days later, after scouring Paris for boots, he was wearing nondescript khaki, a semi-military personage.

While he waited, Myers networked busily. He called on the philosopher Henri Bergson and the neurologist Jules Dejerine. He visited several Allied hospitals, including one at the Trianon Palace at Versailles and another at the Hôtel Claridge off the Champs-Elysées, run by a group of women doctors: 'the body snatchers as they are irreverently called – so successful have they been at getting patients by driving their cars near the front'. He also called in on the Red Cross unit staying in the Astoria Hotel, where he found a Cambridge contemporary, the famous neurosurgeon Percy Sargent, and his medical partner, the London neurologist Gordon Holmes.

On 24 October the hospital's 'kit' was taken to the station at Paris; it filled eleven trucks. But the unit's destination remained unclear: having no tents, it needed a building to accommodate it and none could be found in Abbeville or Boulogne. So the duchess's energy and influence was put to work and finally, on 28 October, the doctors and nurses set off. After a long train journey, followed by a short tram ride, they reached their destination – the Casino in Le Touquet.

For the next week, the team was busy establishing the hospital and persuading the authorities to send them patients. The Casino, which the duchess had managed to prise away from her society rival Lady Dudley, made a magnificent hospital but it took time to persuade the military to make use of it. The chain of evacuation ran from the front line to Saint-Omer and thence to Boulogne, and lastly to Étaples, a further fifteen miles down the Channel coast. Finally, on the evening of 4 November 1914, just as the war was settling down to trench warfare, the first cases began to arrive at Étaples station, from where they were ferried in cars and ambulances to Le Touquet. Myers had been appointed registrar of the hospital and took great care making sure that patients' records and case histories accompanied them; finding the army's system inadequate he devised his own card index (with stationery he ordered from England) and noted proudly that his system was commended by inspecting medical brass. In his spare time he took long bracing walks across the sand dunes and played golf – his play beginning to improve when his own clubs arrived after a fortnight. His wife Edith even came on a visit.[4]

But Myers quickly grew frustrated. Though magnificently equipped, the hospital was amateurishly run. Society figures came and went, almost at will. The duchess, who was accompanied everywhere by her Irish wolfhound, believed that 'one of the best things we could do was to raise the morale of the soldiers, and when a convoy of patients arrived she and the other ladies always dressed up in full evening dress, with diamond tiaras and everything, whatever the time of day', as one nurse later recalled. 'They would parade themselves and stand at the entrance to take the names of the men. They used to set the gramophone going too, so that they would have a welcome. They meant very well, but it did look funny, these ladies all dressed up and the men, all muddy on the stretchers, looking at them as if they couldn't believe their eyes. I remember one man saying, "We thought we were going to Hell and now it seems we are in Heaven!"'[5]

Among the staff, however, the atmosphere was poisonous, with a guerrilla war raging between the duchess and different factions among the doctors. The medical chief, a regular army officer and holder of the Victoria Cross, left after a fortnight. 'A very excellent fellow', Myers wrote. 'Not much used to or fond of "ladies' society".' Myers' own efforts to find a role were constantly thwarted; he was several times given patients by one doctor, only to have them taken away by another. But he did manage to work with one interesting case.

Myers was already aware that the war was producing strange 'nerve cases'. While visiting Jules Dejerine at the Salpêtrière in Paris, he had been shown soldiers who had become dumb or partially paralysed, and in a hospital in Boulogne a Cambridge colleague had demonstrated to him a 'queer case of aphasia', or lack of speech. Then, on the very first trainload of wounded to reach Le Touquet, Myers had noticed a soldier who had been caught by shellfire while entangled on barbed wire. He was now suffering from loss of memory and impaired vision. Myers spent ten days testing the man's senses and observing his gradual recovery, but an attempt to use hypnosis to recover his memory failed. Early in December, Myers was told that most of the staff would be returning to England for Christmas and that he, being Jewish, was 'the most appropriate man to be on duty that day'. He was annoyed: 'So one's race is in everyone's mouth', he wrote. But being on Christmas duty proved a blessing: Myers used the time to observe a second patient, a 25-year-old corporal who had been buried for eighteen hours when a shell blew in the trench he was occupying. This man, too, had almost completely lost his memory, but this time Myers was able to restore it through hypnosis:

> He remembers that he had been two days in the reserve trenches before he was sent on Dec 7th to the firing line. He says, 'The explosion lifted us up and dropped us again. It seemed as if the ground underneath had been taken away. I was lying on my side, resting on my hand, when the shell came. I got my right hand loose, but my left was fixed behind a piece of fallen timber. At last I dropped off to sleep and had funny dreams of things at home. I haven't been able to work out why I should dream of the young lady playing the piano. I don't know her name and don't think I have seen her above twice.'[6]

Excited by the psychological questions this case raised, Myers was now determined to escape from the seaside pantomime. In the New Year, he began spending much of his time in Boulogne, fifteen miles up the coast, where Gordon Holmes and Percy Sargent, the neurologist and neurosurgeon he had met in Paris, were now running a hospital for soldiers with head wounds. They were medical heavyweights, respected by the military authorities. Holmes, a tall, reserved, irascible Anglo-Irishman of thirty-eight, was one of the coming men at Queen Square, already famous for his work on brain structure; and Sargent, a witty, urbane Englishman, was one of the finest neurosurgeons in the country. Through them, Myers came into contact with the Harley Street consultants the army now relied on.

Thanks to one such grandee, an opportunity to do 'psychological' work finally arose. On 30 January 1915, Myers had a chat in Boulogne with Lieutenant-Colonel William Aldren Turner, an experienced Queen Square neurologist, who had been sent out to France by the War Office to report on the high incidence of breakdowns suffered by British troops in the early months of the war. Aldren Turner had found that the very high number of these cases – some seven to ten per cent of all officers and three to four per cent of all ranks were being sent home suffering from nervous disorders – had been caused by the heavy fighting in October, in which the British Expeditionary Force had helped the French to stop the German advance. Men had become paralysed under shellfire or reduced to collapse by exhaustion and strain. However, having organised for such cases to be sent back to specialist hospitals in England, Aldren Turner was now anxious to get back to London himself. 'I spoke to him about the difficulties of psychological work at Le Touquet and hinted at collaboration', Myers noted. 'Then I meet Holmes who spontaneously says, "How I wish you would join us in undertaking the psychological work in the district." So there is a prospect of change.'[7]

Eager to realise that prospect, Myers (who was due a week's leave) hastened to London, armed with a letter from Sargent, which gained him an interview at the War Office with Sir Alfred Keogh, the army's Director General of Medical Services. He also checked up on his two 'psychological' patients who had been evacuated from France and paid a short visit to Cambridge to see his family. But when he returned to France, nothing seemed to have changed. Back in Le Touquet, Myers fretted while Aldren Turner continued to lobby on his behalf. Weeks passed.[8]

Then two events occurred which moved things on. First, a paper Myers had written about the psychological cases he had observed appeared in the *British Medical Journal*: in it the term 'shell shock' appeared for the first time in the medical literature. Then, on 27 February 1915, Myers heard that he had been made a fellow of the Royal Society, for his work on ethnic music. 'Great excitement and entertainment among my colleagues [in Le Touquet]', he noted. 'They appear to have no notion that it could possibly be! Chairing into the dining room. Champagne all round.'[9]

Myers had become an expert. On 16 March he resigned from the Duchess of Westminster's hospital and began helping Aldren Turner with the cases he was seeing. A fortnight later his appointment as

'Specialist in Nerve Shock' to the British Army, with the temporary rank of major, was announced. For an expert in ethnic music, who had scarcely practised as a doctor and had no experience as a neurologist or psychiatrist, it was quite an achievement.

Why was Myers appointed? Why did the army not have its own experts in place to deal with nervous and mental disorders?

Part of the answer lies in the nature and traditions of military culture. In the British regular army, doctors enjoyed low social status and were tolerated so long as they understood that their primary function was to keep soldiers fit and not go looking for complaints. The men themselves were mostly young and healthy; military psychiatry in peacetime was largely concerned with alcoholism, syphilis (especially in officers) and occasional mental health issues, particularly what we now call schizophrenia. In wartime, it was altogether different. The more knowledgeable military doctors were aware of the literature of the past: they knew, perhaps, that during the Crimean War of 1851–4 a condition known as 'palpitation' had been common among British soldiers and that many of the Union troops in the American Civil War had developed symptoms of 'nostalgia' or homesickness. Some doctors would also have taken part in the long debate about 'disordered action of the heart', a cardiac condition seemingly caused by marching or overexertion, which went on within the British military establishment for decades prior to 1914, without ever being resolved.[10]

In the late nineteenth century, as civilian medicine became more complicated and new disciplines such as neurology began to develop, the medicine of war grew more sophisticated. On the one hand, the battlefield could provide human case material for researchers who would otherwise have had to work on animals: head wounds incurred during the Franco-Prussian War of 1870–1 taught German doctors a good deal about the localisation of functions in the brain. On the other hand, arguments within civilian medicine – for example, over the effects of 'railway spine' or the nature of neurasthenia – began to enter the military world, particularly when, as in the Anglo-Boer War of 1899–1902, civilian doctors were brought in to help their military colleagues.

The disastrous defeats which marked the early stages of the British war in South Africa produced several studies of the after-effects of fear. In 1900 Dr Morton Finucane reported on sixty cases shipped back to England and treated at a hospital in Aldershot. His patients had quickly recovered from the light wounds they had suffered at Spion Kop and Colenso, and had no signs of physical injury to the nervous system, but the emotional effects of the 'shock and panic' they had felt in battle were still with them in the form of disorders of speech and memory or physical wasting. These cases of 'functional impairment of nerve sense and motor power, associated with psychical symptoms', were, Finucane argued, 'akin to nervous shock or those observed after railway accidents'. He predicted that it would be necessary to invalid 'a large body of our best and most experienced soldiers out of the service, as being unfitted for future service as soldiers' and added – with rare outspokenness for a military doctor – that 'badly conceived projects by generals and commanding officers causing panic and disaster may then be found to be largely responsible for the development of such nervous cases quite apart from surgical injuries'.[11]

Over the next decade, the British Army was preoccupied with the medical lessons of the Boer War, which mainly related to sanitation and surgery. But experts who monitored reports of foreign wars, such as that between Russia and Japan in 1904, knew what to expect in the future. 'The conditions of modern warfare, calling large numbers of men into action, the tremendous endurance – physical and mental – required, and the widely destructive effect of modern artillery fire will undoubtedly make their influence felt in a future war,' Dr A. G. Kay wrote in the Royal Army Medical Corps' *Journal* in 1912, 'and we shall have to deal with a larger percentage of mental disease than hitherto.' Dr Kay also emphasised the importance of testing the mental capacity of every recruit before enlistment and the need to keep mental defectives out of the army.[12]

Yet these articles had little effect on policy. At the International Medical Congress held in London in 1913, military doctors gave papers on a wide range of subjects but nothing was said about nerves; officially, they did not exist in the armed forces. More importantly when the 'Old Army' had to be rapidly expanded after the outbreak of war in August 1914, no real attempt was made to assess even the physical health of those who rushed to enlist. 'The whole country', one medical examiner later recalled, 'was simply seething with recruits. They were medically examined . . . in the most haphazard manner. 20 to 30 per cent of the men were never medically examined at all. I know of one doctor who medically examined 400 men per day for ten days and he didn't work 24 hours a day.' Large numbers of people joined up who were quite unfit for service life, let alone trench warfare.[13]

Charles Myers stopped keeping a diary on 11 May 1915. This means that, while we can follow in great detail the process by which he introduced the term 'shell shock' to medicine and the assiduous networking by means of which he extricated himself from the Duchess of Westminster's hospital, his progress becomes much harder to chart once he had an important role to play in the war. But a general picture emerges from *Shell Shock in France 1914–18*, the book he published two decades later, from his medical papers, and from the official record.

Frustration, primarily with the military hierarchy, is the dominant tone; Myers referred later to mistakes – 'errors of commission, omission, and, especially of wasteful procrastination' – and added that he had decided not to mention 'certain exceptional personal difficulties with which I was confronted in one quarter in France'. It is not clear what the latter were – perhaps the anti-Semitism that was endemic in the British Army at that time.[14]

The truth, however, is more complex. Having elbowed his way into his position, Myers found himself uncomfortable with the demands that it made on him. Firstly, he was ill at ease with the military. Although he got on well with the very senior people – the urbane, intelligent Sir Alfred Keogh in London and the dapper, sardonic Sir Arthur Sloggett, his counterpart in France – Myers always remained culturally at odds with the military: an academic, intellectual and Jew in an alien world. It was probably the reluctance of military doctors to take decisions for themselves, their acceptance of the need to subordinate clinical judgements to the needs of their military masters, and their insistence on observing the regulations, that he found hardest to accept.[15]

Myers' task when he started in France was 'to arrange the dispatch of mental and nervous cases from France to England', but his role soon expanded into giving advice in courts martial, handling psychiatric and neurological cases for which he had little training, and trying to overhaul the administrative procedures of army medicine. The civilian/military divide was particularly wide when it came to courts martial. During the Great War the British Army executed 307 men for desertion, cowardice and other offences – not a few of whom were suffering from mental disorders – while saving a further 2,700 from the death penalty. Under the regular army's code, which survived intact for most of the war, a commanding officer's right to execute soldiers for such graves breaches of military obligation was regarded as fundamental to the maintenance of strict discipline. 'How can we ever win if this plea [for mitigation of the death sentence] is allowed', Sir Douglas Haig wrote on one soldier's file. Similarly, the decision whether to carry out a death sentence was usually mainly based on the state of discipline in a soldier's unit, and senior officers were more likely to be swayed towards leniency by the man's service record than by his medical history. As a result, Myers found that 'from the military point of view a deserter was either "insane" and destined for the "mad house" or responsible and should be shot'. The first case he saw was a young man charged with desertion who was not intelligent enough to grasp the seriousness of the charge he faced. Myers had him sent home, labelled 'insane', only for him to be returned to France 'with a report that no signs of "insanity" were discoverable in him'. Gradually, Myers found other ways to save soldiers who were not fully responsible for their actions from the firing squad, though how successful he was will never be known.[16]

Myers met with similar inflexibility when he tried to improve the way the army handled nervous and 'mental' patients in France. When he began his job, cases of undoubted insanity, epilepsy, and of 'shell shock' that were too severe to be left in general wards, as well as soldiers who had committed a military offence and were being held for psychiatric assessment, were all flung together in a 'dismal, ill-ventilated and overcrowded' mental ward on the attic floor of a former hotel in Boulogne. Myers felt it was quite wrong to put men who were only slightly damaged among 'the acutely demented melancholic, maniacal, delusional or suicidal patients': 'naturally they feared that they, too, were destined for a "lunatic asylum"'. He managed to get the lighter cases taken for walks along the seafront but his efforts to get a new facility created were less successful. The administrative officers at the various base hospitals were reluctant to provide special accommodation for 'mental patients' whom they felt should be sent home to England. Myers also tried to create ways in which soldiers who were just temporarily shattered could be found things to do in France or returned to the front – again he met resistance. 'We can't be lumbered with lunatics in Army areas', was the general RAMC view.[17]

For all the obstacles which the military put in Myers' way, however, there was perhaps a more important element in his frustration: he was out of his clinical depth. Having introduced the term 'shell shock' into the medical literature, Myers did not clarify what it meant, either because of lack of experience or because he was himself unsure. Over the next year, he published two more 'contributions to the study of shell shock', both cautious scientific papers. The first dealt with five cases he had treated with hypnosis, and although the results recorded were mixed, Myers concluded: 'No one who has witnessed the unfeigned delight with which these patients, on waking from hypnosis, hail their recovery from such disorders can have any hesitation as to the impetus thus given towards a final cure. More especially is this the case with the recovery of lost memories . . . striking changes in temperament, thought, and behaviour . . . follow on recovery from the amnesia.' In fact he had not found it easy to use hypnosis and by doing so he had aroused fierce opposition from the army medical authorities. The second paper discussed disorders of the skin which some soldiers had developed; most of the cases described had been shelled. But neither of these narrowly focused, academic papers provided any sort of overview laying down guidelines for the diagnosis and treatment of shell shock.[18]

It was left to others to try to resolve the mystery. All over Europe doctors wondered whether modern weapons, particularly the development of much more sophisticated artillery using smokeless high explosive, had introduced new nervous complications to the battlefield. 'I wish you could be here,' Sir William Osler, the professor of medicine at Oxford wrote to a colleague in July 1915, 'in this orgie [sic] of neuroses and psychoses and gaits and paralyses. I cannot imagine what has got into the central nervous system of the men . . . Hysterical dumbness, deafness, blindness, anaesthesia galore. I suppose it was the shock and the strain but I wonder if it was ever thus in previous wars.'[19]

'Shell shock' soon became a catch-all term which covered at least four different categories of casualty: soldiers who had been shelled, had no lesions, yet were nonetheless shaken and needed time to recover; men worn down by the strain of prolonged warfare; men traumatised by the horrors they had witnessed on the battlefield; and men immobilised by fear. Some experienced clinicians were aware of these distinctions from the start. According to William Aldren Turner, the Queen Square neurologist whom Myers had replaced in France in May 1915, it was soon recognised that 'one type of case was due to the explosion of big shells in the immediate vicinity of the patient, who did not himself receive any detectable physical injury or wound'. However, he added, 'intermingled with cases of this nature, cases of a general neurasthenic character were found whose symptoms were attributable to exhaustion of the nervous system, induced by physical strain, sleeplessness and other stressful conditions associated with the campaign'. Aldren Turner therefore did not use the term 'shell shock'; he preferred 'battle shock'.[20]

There was no uniformity of opinion among doctors. Some of their sense of confusion is conveyed in a long paper which Dr David Forsyth, an experienced London paediatrician with psychoanalytic leanings, wrote at the end of 1915. He was emphatic that 'These traumatic cases are certainly not the neurasthenia of civil practice':

> The symptoms themselves are hardly ever the same twice over. To enumerate the commonest the patients may appear obviously shaken in nerves, jumpy, and easily alarmed, with a tense worried or harassed expression; or they are dazed or stunned, or even stuporose. They probably feel physically exhausted; almost certainly they sleep badly, starting up, perhaps several times in a night, from frightful dreams of the horrors they have witnessed or undergone. They may be emotional, depressed, reserved, or irritable, and many of them are sexually impotent. They commonly complain of headache, perhaps of dyspepsia or other pains, giddiness, buzzing in the head, palpitation. Their memories fail them repeatedly, especially over proper nouns, and their power of concentration of attention is feeble. They may present some functional disturbances of common sensation, or may be deaf and blind. Most of them are tremulous; many twitch involuntarily, especially in the face. Some stutter; some are aphonic; mutism is the confection of others. Not a few are paralysed in leg or arm.

Many, though, had no particular symptoms and were simply suffering from nervous exhaustion. Forsyth concluded that 'by far the intensest strain is shellfire' but emphasised the 'diversity of effect on different individuals exposed to the same strain'. It was evident 'that the intensity of the shock is to be measured in terms not of the trauma but of the sensitiveness of the individual. Indeed . . . the mere anticipation of active warfare may provoke a neurosis.'[21]

Paradoxically, the doctors' confusions over shell shock – their uncertainty as to whether it was something new in war – only served to increase the public's interest in the subject. Shell shock quickly established itself in the public imagination as a key aspect of the war; it was featured in the newspapers and debated in Parliament. The fact that so many patients were officers – one in six, where there was one officer to every thirty men on the front – helped to make it respectable; there was no talk of degenerates or weaklings now. As early as 1914, public busybodies were concerned about shell shock. Dr Maurice Wright of the Psycho-Medical Society had urged the War Office to set up specialist hospitals. Lord Knutsford, the great hospital administrator and publicist, had taken up the cause, and Lord Northcliffe's brother had offered his grand house in London as a hospital for shell-shocked officers.

Press coverage was mostly vague about what shell shock actually was; phrases such as 'wounds of consciousness' or 'the wounded mind' were frequently used. An advertising campaign which ran in the popular papers in the autumn of 1915 claimed that the nerve tonic Phosferine provided 'the extra nerve force to overcome the bodily discomforts, the brain fatigue experienced under shellfire'. The general impression was that this was a new and dangerous malady, in which bursting shells caused serious disturbance of the nervous system. The fact that the lay conception of the term was so loose and ill-informed had its uses, however, for it enabled the general public to cover behaviour ranging from cowardice to malingering to wounding or insanity and psychosis under a single term. For the relatives of a soldier who had broken down or been unable to face the rigours of the army, 'shell shock' was a great relief. As was later pointed out, while the public was determined that every man capable of fighting should be sent to the front, there was also great solicitude for the wounded and 'this class of case aroused more general interest and attention and sympathy than any other'.[22]

How did public sympathy affect official policy? One important witness believed that 'the contested domain' of shell shock was 'in a great part grabbed by an outsider – the general community; which largely set . . . the policy to which, in practice at least, the medical service and profession [had] to conform'. In some respects it made very little difference: concerns voiced in public and in Parliament did not stop the military authorities in France from executing men who were clearly suffering from shell shock throughout the war. The War Office in London, on the other hand, was more responsive. Sir Alfred Keogh was quick to organise a network of hospitals for shell-shocked soldiers and to take over public asylums for soldiers with mental disorders. He also created a 'Shell Shock Committee', made up of prominent doctors, such as the Queen Square neurologists William Aldren Turner, Farquhar Buzzard and Dr Maurice Craig, Rivers' old colleague at Bethlem and now a prominent psychiatrist. Craig and Buzzard would later be credited with persuading the War Office to 'accept the position that the mind was vulnerable as well as the body, that so-called "shell shock" formed and must form a part of the casualties of modern warfare, and that such casualties required very special treatment and a very special organisation'.[23]

Although Keogh was not on the Army Council, he undoubtedly had a hand in its decision in late 1915 to recognise shell shock as a category in military medicine – that is, to acknowledge, for the first time, the existence of a grey area between madness and cowardice. In doing so, however, it tried to maintain the traditional military distinction between 'battle casualties' and sickness: between wounds, which carried honour and dignity, and simple breakdown, which did not, with 'enemy action' as the distinguishing criterion. The crucial question would be whether or not the soldier had been under enemy shellfire. The army in France was therefore instructed that 'Shell shock and shell concussion cases should have the letter W prefixed to the report of the casualty if it was due to the enemy; in that case the patient would be entitled to rank as "wounded" and to wear on his arm a "wound stripe".' If, however, the man's breakdown did not follow a shell explosion, it was not thought to be 'due to the enemy', he was labelled 'Shell shock S' (for sickness); he was not then entitled to a wound stripe or a pension.[24]

Myers was not consulted about this decision, the effect of which was further to exacerbate an already confused situation. Depending on external circumstances, a shell-shocked soldier might earn a wound stripe and a pension, be shot for cowardice, or simply be told to pull himself together by his medical officer and sent back to duty. Front-line doctors often ignored the Army Council's ruling, labelling patients 'Mental', 'Insane' or even 'GOK' (God Only Knows) before sending them to the base. But soldiers were quick to respond to the new situation and seize on the advantages of shell shock. 'We have seen too many dirty sneaks go down the line under the term shell shock', medical officers complained to Myers, 'to feel any great sympathy with the condition.' The word had become a 'parrot cry' on the tongue of all officers and men, he was told, the 'invariable answer' that soldiers evacuated to aid posts gave to doctors' enquiries. 'Shell shock should be abolished', Myers was informed. 'The men have got to know the term and will tell you quite glibly that they are suffering from shell shock when really a very different description might be applied to their condition.'[25]

Myers agreed: he had 'seen too many men at Base Hospitals and Casualty Clearing Stations boasting that they were "suffering from shell shock, sir" when there was nothing appreciably wrong with them save funk'. 'It had', he wrote later, 'proved impossible to legislate for the bad, without doing injustice to the good, soldier.' Some men were given an easy option out of the trenches, and were taking it, whereas others who were genuinely suffering, were being denied proper treatment. In June 1916, Myers – the man who had introduced the term 'shell shock' into the medical literature – proposed that it should be abandoned by the army and replaced by two conditions called 'concussion' and 'nervous shock', combined with proper facilities for forward treatment of such cases at the front. But there was little official response, and the misleading categories of 'Shell shock W' and 'Shell shock S' continued to be used.[26]

All this time, the medical arguments had continued, in London, Berlin, and Paris. When Britain's leading nerve doctors gathered at the Royal Society of Medicine in London on 4 January 1916, they agreed on one thing – that 'shell shock' was an unfortunate term. It covered, declared Henry Head, 'a heterogeneous collection of different nervous affections from concussion to sheer funk, which have merely this in common that nervous control has at last given way'. He thought it would be just as reasonable 'to sweep up the various fruits which fall from the trees in a strong wind and then to discuss them without first stating that some fell from an apple and some from a pear tree'. Paradoxically, the fact that most soldiers recovered had made it very difficult to get hold of any samples of brain tissue which might provide some explanation of what was going on. The main speaker, the pathologist Frederick Mott, thought that the physiological effects of shelling, psychological trauma and exhaustion were interwoven. He made no attempt to solve the puzzle and simply described in some detail the strange symptoms, such as disturbed dreams, which his patients showed. His vagueness prompted the Queen Square neurologist James Collier forcibly to point out that 'the conceptions of concussion and psychic trauma were antagonistic the one to the other'. Collier thought that 'the real causative agent' was not the shelling itself but the psychological aftermath.[27]

These views were, of course derived from experience with cases evacuated to England, as well as from arguments simultaneously going on in France and Germany. The discussion moved on in June 1916 when the first authoritative study of shell shock in British troops in France appeared. Harold Wiltshire, an experienced London physician, had seen 150 cases during a year at a base hospital in Rouen, and had compared notes with doctors at the front. His conclusions were stark: 'The condition of shell shock was of extraordinary rarity among the wounded . . . the vast majority of these cases, if not all, were due to psychic shock, and not to physical shock.' 'Shell shock' was thus a meaningless term. In sharp contrast to Dr Forsyth, Wiltshire was emphatic that the 'functional nervous affections of modern warfare are essentially the same as the functional nervous affections of civil life'.[28]

The following month, the British launched a major offensive in France. They had chosen to attack in the chalky country surrounding the River Somme.

Private Arthur Hubbard joined the army in May 1916; four weeks later he found himself in France. 'I am with the best of fellows', he wrote to his sister. 'We shall all return back safely together and before this year is through.' But he found the conditions closer to the line not to his liking and soon began to complain about rain, mud, lice, rats and

the 'very tedious work' he was required to do. He also began to envy
his family at home 'sitting round the table about 8.30 enjoying a good
breakfast and me miles away in this miserable place which is being
and has been blown to hell by the Huns'. 'I don't feel inclined to tell
you a pack of lies', he wrote in late June. 'If the truth was told a bit
more often I don't suppose the war would be on now, when you land
over here, they have got you tight and treat you as they think.'

Two days later, on 1 July 1916, Hubbard's unit, the 1/London
Scottish, took part in the great attack on the Somme:

> We had strict orders not to take prisoners, no matter if wounded. My
> first job when I had finished cutting their wire away [was] to empty
> my magazine on 3 Germans that came out of their deep dugouts,
> bleeding badly, and put them out of their misery. They cried for mercy,
> but I had my orders, they had no feeling whatever for us poor chaps
> . . . it makes my head jump to think about it.

Hubbard managed to fight as far as the fourth line of trenches, but
by 3.30 that afternoon practically his whole battalion had been wiped
out by German artillery. He was buried alive, dug himself out, and
during the subsequent retreat was almost killed by machine-gun fire.
He collapsed and was evacuated. Six days later, he was writing to his
relatives from the East Suffolk and Ipswich Hospital to explain why
he was no longer in France.[29]

Private Hubbard had taken part in the worst day in British military
history, in which 57,000 men were lost, a third of them killed. Although
so much has been written about the first day of the Somme, the battle
that began in July was to go on until late October, and it was the long
grinding attritional battles that followed which did most of the damage
in terms of shell shock, and posed the greatest challenge to Charles
Myers. Prolonged fighting and heavy bombardments were, in the
judgement of one experienced front-line doctor, the main factors
causing soldiers to break down:

> The type of warfare practised during the Somme battles of 1916
> provided ideal conditions for the development of these disorders. The
> 'artillery preparation' of the attacking force called for an 'artillery reply'
> from the opposite side. The duel frequently lasted several hours or

even days and during this period of waiting the nerves of all were on edge. Then, after the attack, came the reckoning of the losses among comrades and it was not unlikely that, owing to the call for troops, the whole acute process might be repeated. Little by little men became worn down by such experiences and despite their best efforts the time would come when it was impossible to keep their thoughts from preying on the ordeals and sights of the battlefield. In such instances a break-down occurred slowly; a gradual change would be noticed in the demeanour and behaviour of the patient and he would eventually reach hospital with the report that he was 'quite useless in the line'.[30]

The Somme changed things. Shell shock, hitherto a minor medical problem, now became a severe drain on military manpower. 'In the first few weeks of July 1916,' according to the British official history, 'several thousand soldiers were rapidly passed out of the battle zone on account of nervous disorders and many of them were evacuated to England.' Between 26 July and 11 August, 2 Division, for example, had 501 cases of 'shell shock wounded' as compared to 2,400 wounded. The number of men returned as 'shell shock battle casualties' tripled in the last six months of 1916 to some 16,000 cases. As a result, a battle of wills quickly developed between soldiers, looking for an escape from the hell of the front line, and the military authorities. On 2 August 1916, by which time some 200,000 British soldiers had been killed or wounded, General Headquarters called for greater 'economy in men and reserves'. Soon after, the Adjutant General and the medical hierarchy took steps to clamp down on evacuations for shell shock. Some doctors recognised that shell shock was an inevitable conse-quence of modern warfare. Others did not.[31]

Colonel Bruce Skinner, the energetic and outspoken Deputy Director of Medical Services for III Corps, first began to be concerned about shell shock in early August 1916. Noticing that twenty-five men from 57 Siege Battalion, an artillery unit which was in an exposed position in the line, had been sent to a field ambulance, he at once visited the front-line hospital and concluded that the men were 'suffering from trepidation rather than shell shock'. He then arranged for the unit to be moved back to a quieter spot at night, so that its gunners could get some sleep, but he also summoned the regimental medical officer who had made the diagnosis:

He agreed with me that the cases were due to trepidation rather than shell shock. I told him he must use NYD [not yet diagnosed] if he had occasion to send a man to [a field ambulance] for shell shock, discontinue the use of the term shell shock and use his influence to stop the idea that this battery is being exposed needlessly. He must tell them the general has good tactical reasons for the position and they must do their duty to their country; his influence must be exercised to prevent this rot 'spreading'.

Skinner was not satisfied when the doctor 'promised to help put things right, if he could': 'I told him there must be no reservation, or I would have to place an MO [medical officer] there who would see it done.'[32]

Skinner made it his business to implement the orders from above calling for economy in men and reserves. 'The fashion of running back as having shell shock directly a shell explodes must be checked', he wrote on 12 August, after arguing with another front-line doctor 'who considers shell shock to be the equivalent of a wound'. On 23 August, all the corps medical chiefs were summoned to a conference with Sir Arthur Sloggett, the Director of Medical Services in the British Expeditionary Force who expressed concern about the numbers of shell shock cases being sent down, often without seeing doctors first, 'who had nothing the matter with them'. 'The number of cases arriving at CCSes [Casually Clearing Stations] with a tally marked shell shock by an orderly must be somehow decreased.'[33]

Skinner hotly denied that such things happened in his unit – 'if men got down by improper channels that was a case of desertion and one for the [military police]' he insisted – but nonetheless summoned all the doctors under his command to make sure the policy was implemented. As far as Skinner was concerned, shell shock was a purely physical affliction, caused by contact with shells. But, even with that distinction, he found his policy hard to implement. On 29 August he saw a lieutenant in the Black Watch who had been buried by a shell. 'Judging by what the officer told me', Skinner concluded, 'he was naturally a little shaky after being buried. He has no contusion or shell shock or other adverse symptom. It is difficult to know what to do in a case like this, as the man isn't ill, though in the first instance he had some temporary disturbance of pulse rate. Such cases being returned as "wounded" put a discount on the distinction.'

In his diagnosis, Skinner relied on his own judgement, making rapid impressionistic assessments based on a soldier's service record. Thus, on 7 September he ordered that 'several so-called shell shocks, most of whom were perfectly well', be sent back to duty; but a week later, he accepted that a group of men, including a sergeant major who 'had gone through very severe strain at Flers and at Strongpoint beyond', had been correctly diagnosed and should be evacuated from the line.

Skinner and his doctors agreed on one thing, however: the 'invention of shell shock' had been a big mistake. When the 'shell shock problem' was discussed at a conference on 5 October, 'it was agreed that there is much abuse of the term in order to get out of the front line and that the term should be abolished'. Skinner himself would play an important role in implementing this new line because on 1 November 1916 he was appointed director of medical services for 5th Army, with the rank of surgeon general.[34]

Thus by the end of the year there was general agreement among the British military hierarchy in France, firstly, that men must not be allowed to get themselves away from the front by using shell shock as their route of escape and, secondly, that the term shell shock itself needed to be scrapped. But what should take its place?[35]

It was against this background that Charles Myers' appointment as psychological consultant to the army in August 1916 should be seen. It was a step up of course; it meant promotion to lieutenant-colonel. But, once again, the system expected him to deliver: he was appointed to solve the shell shock problem. Myers' own account is not very forthcoming here, but the records indicate that he was still viewed with some suspicion, notably by his immediate superior in the medical hierarchy. He could not go where he pleased and was only to be summoned to the front when he was felt to be needed.[36]

On 22 September, Myers met with Sir Hubert Gough, the peppery Ulster cavalryman who commanded the Reserve Army. Gough had firm views on shell shock and the doctor's place in the military hierarchy. Earlier in the Somme battle he had personally intervened when a group of men who had taken part in the attack on 1 July and lost most of their officers had reported as shell-shocked. He found it 'inconceivable how men who pledged themselves to fight and uphold the honour of the country could degrade themselves in such a manner and show such an utter want of the manly spirit and courage which at least is expected of every soldier and every Britisher'. And Gough demanded that the doctor who had supported their claim be forcefully brought to understand that 'it is not for a MO to inform a CO that his men are not in a fit state to carry out a military operation'. Gough's thoughts of Lieutenant-Colonel Myers have not survived but we can be sure that he made his views forcefully known during the encounter.[37]

Sometime in the autumn Myers produced his response: a long memorandum arguing that the solution to the problem was to stop evacuating shell shock cases back to England. Instead the British should follow the French example and create special field hospitals near the front line where soldiers could be given treatment quickly, in a military environment, using the appropriate psychotherapeutic measures. Myers tried hard to balance the disciplinary needs of the military with the therapeutic needs of the individual soldier:

Between wilful cowardice, contributory negligence (i.e. want of effort against loss of self-control) and total irresponsibility for the shock, every stage conditioning shell shock may be found. It follows, then, that each case must receive individual attention, based on its own merits . . . The guiding principles of psycho-therapeutic treatment at the earliest stages should consist in the re-education of the patient so as to restore his memory, self-confidence, and self-control. For this restoration of his normal self, a judicious admixture of persuasion, suggestion, explanation and scolding is required.[38]

Soon after, on 12 October 1916, Myers hurried home on seven days' leave on urgent private affairs: his wife Edith had had a miscarriage.[39]

Despite this personal setback, everything seemed to be going Myers' way in France. In November 1916, to his 'great satisfaction', Sir Arthur Sloggett agreed to Myers' proposal for the establishment of four special centres in the rear of army areas where shell shock cases could be held for treatment by specialist doctors.

But, suddenly, things began to go wrong. Early in December, when Myers submitted a long-prepared paper on shell shock for publication in the medical journals, he was told by Sloggett that it was 'inadvisable from a General Staff point of view to publish articles regarding this subject'; instead he should 'confidentially make its contents known to the medical units which would be treating shell shock'. Myers' suggestion that it be printed and circulated within the army was ignored. As Sloggett had spotted, there was certainly an element of academic vanity in Myers' action: he wanted to make his revised views known to his medical colleagues in London, not just to the army.[40]

Then the real blow fell: on 1 January 1917, it was announced that, as part of the new policy towards shell shock, Lieutenant Colonel Gordon Holmes would become consulting neurologist for the northern section of the front and Myers would take this position for the southern section. Where did this leave his previous duties? Angrily Myers asked for confirmation. The official response gave his superior officer obvious satisfaction:

> Lt Col Myers requests information as to whether he still holds the appointment of consulting psychologist and also whether he is to continue to carry out the duties that he has done throughout the BEF in regard to cases of mental disorder. It is considered the answers to Lt Col Myers' questions are in the negative and that he is responsible only for the area assigned to him.[41]

Not only would Myers now have to function as a neurologist; he had lost control of the handling of shell shock in the most important sector of the front and been removed from the hospital for nervous cases in Boulogne which he had worked so hard to improve. Shattered by the turn of events, he retreated to England on sick leave and stayed there for two months.

Reviewing these events two decades later, Myers did not disguise his bitterness towards Gordon Holmes, the man who had eased his path into the military and urged him to take the 'psychological work off his back'. Holmes, he wrote, had no real interest in these cases but had simply got involved with shell shock because his territorial instincts had been aroused:

> Colonel Holmes had previously told me that 'functional' nervous disorders always formed a very large part of the civilian neurologist's practice. Naturally, therefore, he was little disposed to relinquish in army

life what was so important a source of income in time of peace.
Although he confessed that (like most 'pure' neurologists) he took little
interest in such cases.[42]

It was perfectly true that Holmes was renowned for his hostility to
'psychological' approaches to medicine and his contempt for
hysterics. But a more serious charge could be made against him:
that in 1915, when shell shock was seen as a mysterious new disorder,
he had, like most British neurologists, not been prepared to stick
his neck out. Now that neurologists had decided that shell shock
was just a wartime variant on the functional nervous disorders of
peacetime – hysteria and neurasthenia – it was easy enough to pull
rank on Charles Myers.

Yet there was more to Myers' demotion. Shell shock was no longer
an issue of medical speculation, but a matter of military discipline.
The army had to find a way to bring the problem under control – and
ultimately, if possible, to outlaw it altogether – and was therefore
looking for a medical man prepared to lend his authority to such a
process. Holmes, not Myers, was such a man, convinced, as he put it
later, that 'during the Battle of the Somme, a large number of men
deserted from the line on the claim that they had "shell shock" and
it was necessary to prevent that and to keep them within the army
area where they were still under the discipline of the army and could
be reached by their battalion and sent back easily'. Holmes had also
been alarmed by some of Myers' methods of treatment, feeling that
they produced 'a sentimental introspective condition' which was 'decid-
edly opposed to any satisfactory military operations'. His view was
that 'Hysteria spreads by suggestion from one person to another and
has got to be dealt with in no uncertain fashion. Otherwise, the best
army in world finds itself in hospital.'[43]

Sir Arthur Sloggett liked and respected Myers and, rather than
dismissing him altogether, wanted to keep him around in a lesser
capacity, as an academic adviser. And so, after his long break in England,
Myers returned to France and had to watch as others implemented
his plan. He found it painful. 'I have found some of the most successful
MOs among those men who were neither neurologists nor psychiatrists', he wrote to William Rivers in June 1917, as it was clear that
a major British offensive on the Western Front was in the offing. 'The

pure neurologist is hopeless. Yet Holmes is given overall [control of] the now active part of the line for mental and nervous cases.'[44]

In the second half of 1917, the British Army concentrated its main effort in Flanders, fighting thirteen battles between 31 July and 12 November which are now collectively known as 'Passchendaele' and generally regarded as among the most bloody and futile of the war. In the southern sector activity was confined to a much smaller engagement at Cambrai, between 20 and 30 November, in which tanks were used in battle for the first time. Gordon Holmes ran the shell shock arrangements at Passchendaele; Charles Myers at Cambrai.

Myers had got the idea of forward treatment centres from the French. As early as May 1915 concerns at the numbers of functional and psychological cases being referred to base hospitals, and therefore lost to fighting units, had led French neurologists to argue that such disorders were 'perfectly curable at the outset . . . such patients must not be evacuated behind the lines, they must be kept in the militarised zone'. A network of forward treatment centres was then created, usually by the personal initiative of doctors, and in December 1916 the head of one such unit claimed that ninety-one per cent of patients had been successfully treated 'through a simple and energetic psychotherapy and sent back to the front after a few days', adding that electric shocks applied to dysfunctional parts of the body, though 'not essential . . . affects the rapidity of the result'. This success was explained by the general environment of the centres: as two French doctors explained, 'Their only relative degree of comfort, their strict military discipline, their proximity to the front, their remoteness and their inaccessibility to friends and relations render them specially suitable for this form of treatment and ensure much easier and quicker cure than in the interior.' However, these statistics disguised the problems in the French system – most notably that, as time went on, many soldiers were evacuated to base hospitals with the symptoms of physical disorders and there developed psychological problems which delayed their return to the front.[45]

Whatever the imperfections of the French system, however, it seemed to the British greatly to reduce wastage of manpower. But the Adjutant General was only prepared to follow the French example if safeguards against possible abuse of the system by cowardly soldiers and soft-hearted doctors had been built into it; most notably soldiers evacuated from the battlefield to forward treatment centres had to be held there, without receiving treatment, until it had been formally established from their commanding officer that they had been genuinely 'shell-shocked' – that is, been in contact with a shell.[46]

In the first half of 1917 the British Army created four 'shell shock centres', between ten and twenty miles from the front line, at Arques (near Saint-Omer), Frévent (just north of Amiens), Doullens, and Corbie; later, a fifth was added, at Haringhe (or 'Bandaghem'), near Ypres. They were rudimentary affairs, usually consisting of bleak rows of tents pitched in the grounds of a chateau or factory, and were run by a curious assortment of doctors: William Brown, a brilliant academic psychologist and McDougall's first student at Oxford, who had by now worked extensively with shell shock patients in Liverpool, Egypt and London; Frederick Dillon, later a noted London psychotherapist; the neurologist William Johnson, who had won the Military Cross while serving as a regimental medical officer on the Somme; and Dudley Carmalt Jones, the dean of Westminster Hospital Medical School and an old Queen Square friend (and flatmate) of Gordon Holmes. It was as if all the various schools of medicine were being tested against each other. There was not even a standardised diagnosis. Dillon, for example, divided his patients between seventy per cent who displayed direct anxiety or fear characterised by generalised shaking, nervousness, jumpiness, and sometimes dizziness and headache, but who generally recovered in the course of three or four days in the atmosphere of the hospital; twenty per cent who had conversion symptoms such as mutism and paralysis and 'were so easily and rapidly amenable to treatment as to be practically an indifferent factor in the estimation of the case'; and ten per cent split between mental confusion, amnesia or 'fugue' (identity loss which makes the sufferer suddenly disappear from their usual haunts and reappear in some distant place, unable to recount what has happened in between); combined types of cases in which a war neurosis developed in conjunction with an organic disorder or with a previous neurosis. By contrast, Carmalt Jones's main distinction was between cases of rapid onset (seventy per cent) and those of gradual onset (thirty per cent).[47]

There was, however, some common ground when it came to treatment. The doctors agreed that an atmosphere of optimism and expectation of cure, plus the speedy reimposition of military discipline, was important; physical symptoms such as deafness and mutism could then be quickly dispelled; no one should be allowed to think he was an invalid. The most obvious difference was between Brown and Dillon, who both made some use of hypnotism to bring back memories and cure symptoms, and Johnson and Carmalt Jones, who rejected this method. William Brown, who was a skilled and experienced hypnotist, 'a wizard who mesmerises when he pleases', according to one patient, the poet Wilfred Owen, claimed that by recovering traumatic memories quickly he spared patients more suffering further down the line. By contrast, Carmalt Jones distrusted the psychological approach altogether and felt that suggestion 'was used to convey an impression of the occult'; indeed he saw no need for 'systematic mental investigations' and preferred military discipline and gentle encouragement.[48]

Whatever the methods used, these doctors all claimed considerable success in sending about eighty per cent of soldiers back to battle. But a recent examination of the records of one unit suggests their actual success rate was far lower, while there are no reliable figures on the number of men who relapsed. The overall impression is that the treatment centres worked quite well in slack periods and with 'light' cases, such as those described by William Brown:

> The majority of the nerve cases that came down during the first 48 hours after our tanks and infantry went over at Cambrai in November 1917, were very light. They were either old cases of 'shell shock' who had lost their nerve again at the prospect of being heavily shelled, or else men constitutionally weak of nerve and lacking the power to pull themselves together in the face of an emergency. They came down in lorries as walking cases and made a sorry show in the reception room, with their hanging heads and furtive looks. They gave the impression of men who had, at least temporarily, lost their self-respect, many of them were keeping up, with obvious effort, rhythmical tremors which had no doubt been involuntary and irrepressible some hours before, but were now within the field of voluntary control.[49]

Once the battle began in earnest, however, the flood of shell-shocked soldiers quickly overwhelmed the treatment centres, with matters made worse by the need to get every soldier's commanding officer to fill in Army Form 3436, which confirmed that he had actually been shell-shocked. During the Passchendaele fighting the system broke down repeatedly, and many cases had to be evacuated back to England. To reduce admissions, the decision was made not to bother sending men who were exhausted or 'only temporarily shaken' to the forward treatment centres. As Gordon Holmes put it, the majority of 'men temporarily shaken up by the explosion of a shell or buried' required no treatment and were fit for duty after a few days' rest. In other words, by this stage of the war neither simple shell concussion nor plain exhaustion qualified for specialist medical attention any longer. The phenomenon of 'shell shock' was being contained; eventually it would be abolished altogether by the army. At the end of 1917 one of the treatment centres was consequently shut down and turned into a VD clinic.[50]

In the autumn of that year, Charles Myers found a new role. The battles of Arras, Messines and Passchendaele had produced a flood of serious shell shock cases, overwhelming the medical facilities in France. The hope that they could be retained in France proved unfounded. These soldiers had to be evacuated back to Britain and large numbers ended up in general hospitals where, because of lack of knowledge among the medical staff, many of them were invalided from the army. With manpower short and the question of the cost of pensions looming, the official mind was now concerned to stem the further loss of fighting men. An emergency conference was held at the War Office between 15 and 23 October 1917. Myers travelled to London to attend and, while there, was invited by Keogh to become involved in the handling of those shell shock cases back in Britain.[51]

And so, Myers' unhappy period in France finally came to an end.

'The father of a multitude of helpless children'

The war gave William McDougall the opportunity to escape from Oxford. In August 1914 he and his brother Percy, a general practitioner in Manchester, rushed over the Channel and offered their services to the French. While Charles Myers was nursing his frustrations in Le Touquet, McDougall was in the thick of it, as a private in the French Army, 'driving an ambulance and dodging shells on the Western Front'.

Two grisly episodes marked his time in France. On the first occasion, his ambulance picked up a wounded French soldier who had just bayoneted a German. He described to McDougall how 'the German stood erect and, holding in both hands his bleeding bowels as they gushed out through the gaping wound, said in excellent French, "See what you have done to me"'. The Frenchman, as McDougall later reported, 'expected to bear with him to the grave the horror and remorse of that moment'. Then, in February 1915, as the guns boomed and star shells gleamed fitfully along the line of the trenches, McDougall helped to carry into an improvised hospital a strapping young Frenchman from the Midi who ceaselessly murmured, '*Marie, je t'adore.*' A nurse immediately kneeled down beside him, kissed his brow and murmured words of encouragement. Soon afterwards McDougall helped carry the young man to the operating table, but he was terribly wounded and, as the surgeons began to clean him up, the life passed out of him.[1]

In the spring of 1915, McDougall returned to England and applied to the War Office, which by now was short of doctors. His psychological expertise was immediately recognised: he was given the rank of major and put in charge of a ward for nervous cases at Netley. Built in 1870, the Royal Victoria Military Hospital was a monument to the British Empire and to the army that maintained it, a great,

hideous pile of masonry stretching for a quarter of a mile along
Southampton Water. (When the Americans took it over in the next
war they drove jeeps down its corridors.) In 1915 Netley was the clearing
hospital for the rest of Great Britain, its original building now
surrounded by an ugly cluster of huts and tents. Nearly all the wounded
from the Western Front passed through its gates and not a few ended
up in Netley's infamous 'D' Block, the army's main mental hospital.[2]

By now, however, the authorities had come to realise that many
soldiers who seemed 'mental' were in fact just temporarily affected
and so had posted specialist doctors to weed out the cases 'of a neurological character', who were then transferred to the neurological
section, in the main building. McDougall's job was to select from this
flood of mental and nervous cases whatever patients seemed most
susceptible to treatment. 'I was the head of a hospital section full of
"shell shock" cases, a most strange, wonderful and pitiful collection
of nervously disordered soldiers, mostly purely functional', he wrote
later.[3]

McDougall had been fortunate. Whereas Myers was struggling with
the military bureaucracy in France – forced to reconcile the needs of
the military and those of the ordinary soldier, without much time to
relate to individual patients – McDougall was working within the
more enlightened medical environment in Britain and able to get to
know his charges. 'Sympathetic rapport with the patient was the main
thing,' he recalled, 'a very natural and simple human relation':

> I felt like the father of a multitude of helpless children, hopelessly
> stumbling on the brink of hell; and that they for the most part were
> very docile and dependent and grateful. It was a wonderful experience
> for a psychologist; and besides, for the first time in my life . . . I was
> giving my whole time and energy to work that was indisputably worth-
> while.[4]

The war forced William McDougall, the arrogant elitist intellectual
and eugenicist, to mix with and observe his fellow men, in all their
messy forms, and he found the experience both humbling and
enriching.

The case histories McDougall later wrote up movingly record the
effects of the war on individual soldiers. There was, for example, a

young man preparing for the Christian ministry who had volunteered and been sent to the Western Front as a stretcher-bearer, only to find that shell explosions provoked in him uncontrollable fear, which no amount of praying would overcome. Even when removed from the front line, the internal conflict between his fear, his sense of duty, and his faith continued. 'On almost every occasion that I came near him,' McDougall recorded, 'he would draw me aside and beg earnestly to be sent back to the front. He was the victim of a conflict of which there could be no solution so long as the war continued; there would be no peace of mind for him, unless he should be fortunate enough to receive a severe or fatal wound. I have no doubt of his entire sincerity, when he expressed the desire to find death on the battlefield.'[5]

Another Netley patient was an Irish Catholic sergeant, who had been the trusted leader of a group of bombers, all from his home town, whom he had led into many a fight. But 'on one fatal night the squad became engaged in a terrible confused hand-to-hand fight, from which only he emerged alive'. Now this man sat in the hospital in a state of deep depression and every morning he bathed, hoping to wash away the sins of his comrades who had died before they could be absolved of their sins. He was afraid to go on leave, for everyone at home would ask him 'Where are the boys that went with you?'[6]

Some cases had a happy outcome. There was a man who had completely lost his memory after being knocked over by a shell explosion though otherwise he seemed fine. McDougall assured him all would be well and waited for something to happen. A few days later patients arrived from the tropics, wearing sun hats which were hung above their bunks and when the amnesiac soldier happened to see them, he rushed up to McDougall, took him to the ward and pointed excitedly:

I guessed at once that he had seen service in India. I therefore began to repeat to him and to write down before him the names of all the military cantonments in British India, so far as I could remember them. He was immensely interested, his excitement increased, until suddenly his amnesia was gone, as though some barrier in his brain were snapped across; and he poured out a flood of soldier's reminiscences beginning in India and going on to France. No symptom remained, and he went cheerfully back to duty after a further rest.[7]

Many of the patients were classic cases of hysterical dissociation – they had reacted to some awful experience at the front by developing physical symptoms, such as deafness, paralysis or stuttering.

> A sergeant, fighting on Gallipoli, stooped to pick up a bomb which a Turk had hurled at him, intending to hurl it back at the enemy. As he reached for the bomb it exploded. He was not wounded or stunned; but he opened his mouth widely (without doubt as the first step in the natural fear reaction of uttering a cry), and then found that he could not close his mouth or withdraw his tongue, which remained protruding. After some hours his tongue gradually withdrew and his mouth closed; but he was then completely mute; he could not mutter a sound. He remained mute for months, and proved to be a most obstinate case of mutism, defying all my efforts, and only very gradually learning to speak again.[8]

Before the war, French neurologists had made a particular study of the after-effects of traumatic events and, in a series of classic studies, Pierre Janet had put forward the theory of dissociation, according to which the mind bundles away out of consciousness the experiences and emotions it is unable to cope with. In the military context, this meant that the origin of the symptoms that men presented at Netley could be found in the intense emotion of fear caused by shell explosions. Trying to repress and control the fearful emotion at the beginning had brought about a splitting of the mind, or 'dissociation', which expressed itself later as the loss of memory and losses of function also – such as dumbness, deafness, tremulousness and paralysis. The therapeutic rule of thumb was that if the suppressed memories of the incident could be brought to the surface and reintegrated with the conscious mind then the symptoms would disappear.[9]

William McDougall was an accomplished hypnotist, having spent a decade perfecting his technique, and he made extensive use of this technique at Netley. The results he obtained varied. One successful case was a man suffering from 'fugue'. McDougall's patient was a regular soldier, a colour sergeant with a fine service record, who had been riding a motorcycle, carrying a despatch from one part of the front to another; and then suddenly found himself, a few hours later, pushing his machine through the streets of Boulogne, a hundred miles or so from the front. He was utterly bewildered and, to avoid suspicion of desertion, surrendered himself to the military police. No doubt this, and his record, saved him from the firing squad.[10]

The man was admitted to several hospitals, still unable to explain what had happened, until he reached Netley. McDougall tried hypnosis:

> at once the amnesia yielded; the dissociative barrier was overcome, and he continued in the waking state to able to recollect and describe the whole incident; how a shell exploded near him, throwing him down, how he remounted his cycle and set off for the seaport; how he found his way by studying the signposts and asking questions, etc. It was clear that, though his actions had been conscious, intelligent, and purposive, yet his conscious activity was of a restricted kind; he seemed to have had no thought about the consequences of his action, but to have been driven on by the single strong impulse of fear, taking the form of a desire to get away from the danger zone.[11]

Often, however, there were complications, for example, in the case of a sturdy Irishman admitted for confusional 'insanity' ten days after being rendered unconscious by a shell explosion. McDougall found out that the man had survived a drunken father, endured the siege of Ladysmith in the Boer War and never known fear while serving in France. But his battalion had recently taken very heavy losses. As McDougall recounted:

> The night before he was knocked out by the shell, he took part in a night attack. His company was crawling forward in the dark; a shell exploded near him and he put out his hand to touch the comrade beside him and to make sure that he was all right; his groping hand found the stump of his comrade's neck, from which the blood was spurting. This incident seems to have been of the most effect. He says it will be always with him; he often dreams of it; sometime he is aware of the body beside him all night long. After telling me of this, he is no longer troubled by a sense of the presence of the headless comrade. But he still has the sense that something dreadful is about to happen; and he feels fear and wants to run away; what he fears is undefined, but some calamity either to himself or to the whole place. He can't associate with other patients. If he goes to the dining hall he feels as

though he must scream. It is difficult to persuade him to go out of doors; and when he goes out he keeps close to buildings. He is ashamed of these weaknesses.

The condition was obstinate and, since there was no obvious amnesia or repression, McDougall tried indirect suggestion through hypnosis. He found that half an hour's hypnotic sleep, with suggestions of confidence and cheerfulness, made the patient feel all right for three or four days, and he seemed to be steadily getting the better of his nerves. But as soon as he was sent home he relapsed, his vague fears returned, he lost his appetite and was unable to sleep. So McDougall renewed the treatment by direct suggestion, gradually increasing the intervals; and in the course of some two months the man regained his self-confidence.[12]

By a strange chance, one of McDougall's failures has been preserved on film. Before the war Private Percy Meek had been a basket-weaver in Snettisham in Norfolk. During a heavy bombardment in 1916, he had climbed out of his trench, vowing to bring back the trench mortar that was shelling it, and been restrained and taken to a casualty clearing station. By the time he reached Netley two weeks later he had developed tremor of the limbs and could not speak. McDougall managed to communicate with him in writing but progress was so slow that he decided on hypnosis. This revealed that Private Meek was visited every night by the ghost of a German he had killed on the Marne in 1914.

> During the night the figure appears suddenly in the ward, points his rifle at PM, says 'Now I've got you, you can't get away', and fires point blank at him. PM hears the crack of the rifle and sees the ghost sink to the ground. He takes this to be a real ghost come to take his revenge and every night he is terrified anew by this visitor.

McDougall explained this hallucination to the young man as fully as possible and expected him to get better. While Meek stopped seeing the ghost, he did not otherwise improve; his twitching got worse, his muscles more rigid, and he could not even whisper any more. McDougall 'suffered the mortification of seeing the patient regress gradually under my hands, despite all my efforts'.[13]

This failure may have made it easier for the therapist to abandon his patient. In the autumn, McDougall organised a transfer to Oxford, to be near his family, but left Meek behind at Netley. 'Since I had failed so miserably to relieve PM's condition, I thought it best to leave him to the care of other hands', was McDougall's later rationalisation. Private Meek now passed into the care of Dr Arthur Hurst, a neurologist with a reputation for curing hysteria by suggestion, but in this case he too achieved no miracle cure. The patient now regressed completely into childhood, gave up all attempts to walk or speak and spent his time playing with dolls. Private Meek can be seen in Hurst's film *War Neuroses* cowering in a wheelchair, with a teddy bear on his knee, being petted by a nurse. Almost a year later, however, in November 1917, Meek made a spontaneous recovery, after which his memory returned, and he began gradually to regain the use of his limbs. A year later his recovery was complete and he was filmed again, teaching basket-weaving to other patients.[14]

William McDougall had failed Private Meek. How effective, then, was he as a therapist? The evidence suggests that, for all his human sympathy with his patients, he did not have the personality fully to help them. That was certainly the impression the American surgeon Harvey Cushing formed when he visited Oxford in May 1917 and called in on the improvised shell shock hospital, made up of tents and shacks, that McDougall was now running in the gardens of New College. Over lunch, Cushing heard about one of McDougall's patients. A soldier in the regular army, he had served at the front for over two years, survived Mons and the Somme, and been 'over the parapet' some nineteen times. But two months earlier, he and his pals had been caught in a British artillery barrage, and he had seen most of those around him killed, yet he had kept on fighting, capturing some German prisoners in a dugout. He had been about to bring them back when an officer had arrived and shot them all out of hand. Soon after the soldier had begun to twitch and tremble and to weep. 'It's often very wise to let the men talk and weep themselves out', McDougall told Dr Cushing.[15]

In fact, McDougall had found a more fulfilling role – as a mentor to others. He collaborated with Thomas Saxby Good, the medical superintendent of Littlemore Hospital, the asylum outside Oxford, who had been experimenting with psychotherapy since 1911 but had found that his 'preconceived notions of various mental states' left him ill-equipped to handle the complex cases – such as sufferers of fugue, hysterical stupor and loss of memory – that the war was throwing up. Dr Good was bewildered by the way that 'some patients, unable to speak, would write down their experiences, but on regaining their speech would lose their memory; others had memory so reduced that they talked and behaved like children, even as regards feeling and movement; others were even unable to feed themselves, not knowing how to chew or put food in their mouths – reduced in fact to the mental level of a newborn babe'. McDougall taught him to use hypnosis, and soon, '"complexes", "amnesias" and "dissociation" were everyday words' to Good, 'just as previously we had talked of epilepsy, hysteria, neurasthenia and dementia praecox'.[16]

Next the two men addressed a wider problem. Throughout the war, a large number of shell shock patients had been discharged uncured from military hospitals and sent home. Thomas Salmon, an American psychiatrist who toured English hospitals in 1917, described in his report how many soldiers 'enter[ed] the hospitals as "shell shock cases" and c[a]me out as nervous wrecks'; 'mismanaged nervous and mental cases' were, he wrote, 'exposed to misdirected harshness or to equally misdirected sympathy, dealt with at one time as malingerers and at another as sufferers from incurable organic nervous disease', before being discharged. 'To turn adrift thousands of young men who developed the nervous disability through military service and can find in their home town none of the facilities required for their cure' was, Salmon believed, 'indeed a gross injustice'.[17]

McDougall and Good found many uncured cases of shell shock in the hospitals around Oxford and also men living at home or in military camps with uncured symptoms. So they opened an out-patients' clinic at the Radcliffe Infirmary and in 1918 took over Littlemore as a 'neurological' war hospital. They now began to see more complex cases in which pre-war psychological problems, wartime experiences, and difficulties in readjusting to civilian life were combined. McDougall treated a highly educated man whose wartime symptoms returned every time he travelled by train, and discovered, under hypnosis, that a repressed memory of a traumatic journey in a field ambulance lay behind his problem. Another man's symptoms returned when he shaved: he had repressed the memory of an unexploded shell that had fallen while he was shaving at the front. And then there was the case of a sergeant whose shell shock-induced amnesia prevented him from getting a divorce – because he could not remember the circumstances of his wife's infidelity he was forced to go on living with her. All McDougall's efforts to unlock the memory were unsuccessful.[18]

McDougall and Good used two types of therapy – the waking or conscious, and the hypnotic. Waking therapies included suggestion, explanation and re-education with some 'electrical stimulation by faradic current'. The hypnosis included, according to Good, 'specific suggestion under hypnosis, the implanting of general idea of health and peace when the patient was in an hypnotic state, and hypno-analysis, or the regaining of memories in the hypnotic state'. Dr Good noted that during the war most cases seemed susceptible to hypnosis because 'as all classes of the community were under a great emotional stress at this time, the desire for company and mental support was very strong and therefore the herd tendencies were reinforced'; and 'transference was more positive, as the instincts with emotional reactions had thrust themselves nearer consciousness than is usual in the ordinary life of the present day'.[19]

In this favourable climate a miracle healer appeared. Arthur Hadfield was a former student of McDougall's, a missionary's son who had read theology at Oxford, worked as a Congregationalist minister in Edinburgh and served in the navy for several years. He had a powerful messianic personality which lent itself to the technique of hypnotic abreaction. Confident in his own powers of healing, Hadfield boldly attacked the fear buried inside his patients' memories:

The patient, hypnotised, is told that when the physician puts his hand on his forehead he will feel himself back in the trenches, or under fire; that he will visualise all the circumstances when he was blown up or buried, and live again through all the emotional experiences concerned. He is encouraged to describe them as he experiences them. The patient usually becomes disorientated, shows signs of great emotion, trembles, perspires, engages in violent movements, and may be speechless with horror or cry out with fear.

This could be a dangerous technique in the wrong hands, but it worked for Hadfield. He was emphatic that the experiences had to be recalled emotionally: 'Knowing does not cure; it is the release of repressed emotion which is necessary to cure, and that is why the patient is made to live through the whole experience with emotional vividness: knowing only cures in so far as it brings about this emotional readjustment.'

To cope with the huge numbers of patients in Oxford, Hadfield devised a technique known as collective hypnosis, where he would 'subject seventeen to twenty men to suggestions of sleep, comfort and freedom from worry, the men lying on their beds in a semi-darkened room'. Those who did not immediately go to sleep would each be hypnotised by 'fixed gazing', then suitable general suggestions of confidence and reassurance would be given and, finally, individual suggestions appropriate to each man would be added. Hadfield believed his methods worked. He followed up his first hundred patients eighteen months after their discharge into civil life and found that ninety per cent of them were now working full-time at their pre-war occupation; only ten per cent were uncured or only partly cured and receiving pensions.[20]

The work of the Oxford outpatients' clinic impressed even McDougall's bitterest critics. 'The Senior Common Rooms at Oxford had come round to becoming more tolerant of the "new psychology" and its scientific approach,' Hadfield later recalled, 'mainly on account of their therapeutic success in the treatment of shell shock people at their gates.' Several dons, including the philosopher and later vice chancellor, Sir David Ross, even 'sat in' on treatment at Littlemore.[21]

But McDougall himself made little of this achievement, then or later. Was he, perhaps, jealous of the success which Good and Hadfield achieved with methods he had taught them and which, he freely acknowledged, were much less effective in his own hands? He certainly disapproved of some of the more extravagant claims made for hypnotic abreaction. At a seminar on how hypnosis effected cures, in London in February 1920, both he and Charles Myers agreed that the simple release of repressed emotion did not in itself guarantee recovery. The patient had also to be aware of what had happened to him. 'It is the recall of the repressed scene, not the "working out" of the bottled up "irrational energy" . . . which is responsible for the cure', Myers insisted. McDougall agreed: 'The emotional discharge is not necessary to this, though it may play some part.' Neither Myers nor McDougall had been conspicuously good therapists, whereas their opponent in the discussion, William Brown, was generally acknowledged to have been very effective with soldiers. For him, 'The essential thing seems to be the revival of the emotion accompanying the memory.'[22]

There was truth on both sides. During the war there were many cases of simple dissociation, terrified soldiers whom a therapist with a powerful personality such as Brown or Hadfield could easily cure. When, after the war, however, the Oxford clinic began to treat military pensioners and, in a pioneering step, civilians with minor mental health problems, it was not so straightforward. The pensioners were usually chronic cases who had been through many forms of treatment; they resented hypnosis and it required hours of patient psychoanalysis to help them.[23]

It remains something of a mystery why William McDougall, the most prominent psychologist in Britain before the war, did not play a greater role in the wartime discussions and arguments about shell shock. For one thing, he seems to have been inhibited by a lack of clinical experience in this area of medicine; his psychological theories did not lay particular emphasis on the effects of trauma. Then again, although in the 1900s McDougall had taken great pride in bringing the very latest research in physiology into his psychology, he had fallen somewhat behind in the ensuing decade. This mattered, because after 1916 the shell shock discourse entered a new phase. Having previously been both 'physical' and 'psychological' – concerned with the effects of shellfire and with the effects of fear and mental conflicts on the mind – it now entered a third, 'physiological', phase, looking at the body's response to the prolonged strains of warfare. Doctors now pressed into service, not so much Charles Sherrington's work on the interactive action of the nervous system in enabling the entire body to function, but more recent research carried out by the Harvard physiologist, Walter Bradford Cannon. In *Bodily Changes in Pain, Hunger, Fear, and Rage* published in 1915, Cannon demonstrated the role that the human endocrine system – the chemical messengers that communicate between the nervous system and the vital organs – played in mobilising the body for 'fight or flight'. When confronted by danger, Cannon's work showed, evolution had equipped men to mobilise the body to fight or to run away. If, however, a soldier had to remain inert in a trench under shellfire his body had no opportunity to 'work off' the energising substances in the blood.

Another factor was that before the war McDougall had publicly opposed the work of Sigmund Freud and so could hardly use Freudian ideas as an interpretative tool with which to understand the effects of shell shock. But the main reason seems to be that McDougall did not feel himself to have been a success as a therapist. Much of a doctor's aura in this field derived from charismatic authority, a conviction born of successful outcomes – as when the early Freudians claimed that their methods must be right because they worked. For all the successful work done at Littlemore, McDougall was far too honest to make that claim.

In one respect, though, McDougall was more forthcoming. He felt that the war had completely answered the fears expressed in the previous decades that modern urban man might be too soft for modern warfare; the anxieties whether, as McDougall put it, 'in a clash of arms against some more primitive people, we night not be overborne and swept away by sheer lack of nerve, of animal courage; whether our town-bred bespectacled young men, their imaginations quickened by education, and unused to physical hardship, pain and bloodshed, might not shrink and crumble when brought face to face with the horrors of war'. On the contrary, he concluded, 'regiments of cockneys from the London suburbs and of Lancashire lads drawn from the mills and factories of the world's greatest industrial hive, had distinguished themselves by gallantry and by patient courage in the field'.[24]

16

'The pursuit of the Muse'

On 4 August 1914, the day that Britain declared war on Germany, William Rivers was on the high seas, crossing the Indian Ocean on his way to Australia to attend the British Association's 1914 meeting, where he was due to give an address on the culture of the Aborigines. He was too busy correcting the proofs of his *History of Melanesian Society* to pay much attention to a young Cambridge colleague travelling with him. John Layard, the 23-year-old great-nephew of the archaeologist Henry Layard ('Layard of Nineveh'), had come up to King's College, Cambridge to read modern languages in 1909. There he joined the university's Anthropological Club, of which Haddon and Rivers were the leading lights.

Layard remembered Haddon as 'a very nice kindly man with a stutter giving rather incomprehensible lectures on the races' and Rivers as 'a bachelor and to me as an undergraduate an extraordinarily interesting personality [with] a way of enthusing young undergraduates in anything he happened to be interested in'. In his unpublished autobiography Layard recalled that 'Rivers had immense quantities of clever young men around him. He told me once that it was only the affection of young men that kept him alive.' Layard was 'overwhelmed by the aura of this marvellous man' and came to 'worship' him.[1]

Both Haddon and Rivers spotted the young man's potential: would-be anthropologists were thin on the ground. Haddon offered Layard the chance to stay up at Cambridge for a fourth year, studying anthropology, and arranged for his protégé to accompany him to the British Association's 1914 meeting, the plan being that Layard would then travel to New Guinea with Haddon to do fieldwork there. In the end, on the voyage to Australia Layard sailed with Rivers on one ship and Haddon and the other anthropologists followed on another.

When the news came that war had been declared, the German academics on board the Rivers/Layard boat (and due to attend the meeting) suffered some embarrassment; most of the British delegates made immediate arrangements to return home. Rivers' response, however, was to cable his Cambridge college to ask for a term's leave of absence to stay on after the conference, in order to do fieldwork in the New Hebrides. He wanted to follow up reports that a Swiss ethnologist had found a megalithic culture there, a community still erecting temples made of large stones – a living example, perhaps, of the diffusionist thesis. When permission was granted, Rivers invited John Layard to accompany him, rather than go with Haddon to New Guinea. In any case the gunboat in which Haddon had planned to travel was no longer available, as it was required for the war.[2]

Layard and Rivers got a passage to the New Hebrides on a little trading boat, sleeping on benches in the lounge as there were no cabins. Things were more comfortable on the island of Fate where they stayed with Merton King, the British Resident Commissioner, a hospitable man who had been banished to this remote spot for some misdemeanour and spent his days reading *The Times* from cover to cover. On King's advice – and travelling in the 'broken-down Cowes yacht' that he put at their disposal – they made for the small island of Atchin, off Malekula, where the locals were said to have 'refused to be missionised'. Six native sailors in white ducks put the two anthropologists and their stores ashore on the island and they were given the key to an abandoned Roman Catholic mission house, which would be their home.[3]

The Resident Commissioner had omitted to mention that the building had until very recently been occupied by an Irish trader who had made himself so unpopular that the natives had driven him out. Consequently, Rivers and Layard found themselves surrounded by very hostile locals. After a week, however, the natives' curiosity got the better of them and something of a thaw began. 'I think Rivers made the first advances and he was very good – always by touching, he'd touch people', Layard recalled. 'And he managed to get into conversation with a few of the younger natives who knew pidgin English . . . Gradually we made friends.' This edgy relationship is captured in a photograph that shows Rivers standing outside the mission house, surrounded by young men.[4]

But then, to Layard's great surprise, Rivers suddenly disappeared: he had jumped on a motorboat belonging to a group of missionaries that happened to call at Atchin and was taken off to a neighbouring island. He did not return and would never offer a proper explanation for his action. Layard later assumed that Rivers had left because 'he didn't like the discomfort of living a pretty primitive life, living off vegetable food only'; he had, after all, previously done much of his fieldwork from the comfort of the missionaries' schooner *Southern Cross*. But that alone would not explain Rivers' behaviour. Layard would later hint at another reason for the older man's abrupt departure. 'Rivers of course was extremely puritanical', he said in an interview. 'This is the thing nobody realised. People worshipped Rivers because he was so forthcoming, so friendly.' It may be that Layard, who adored Rivers and to whom it came naturally to express intimacy in physical ways, had made overtures to Rivers. And so Rivers had fled.[5]

Layard coped surprisingly well with being abandoned: he made friends with the islanders, learned their language and collected their genealogies. According to his biographer 'he delighted in the life that he led, transcribing their songs, participating in the male initiation ceremonies as an honorary novice, pleasuring in the complex rhythms beaten out by slit-drum orchestras, thrilled by their dancing – in which he once joined, painted black with charcoal and clad only in a penis wrapper'. Layard himself described Atchin as 'my paradise – the one place I'd been really, really happy . . . living with these natives and enormously enjoying life with them'. [6]

At Christmas, Layard accepted the Resident Commissioner's invitation and returned to the main island. There he met Rivers, who had spent his time touring the New Hebrides on the *Southern Cross*, and was now on his way home. He offered no explanation for his departure from Atchin. Rivers reached England in the spring of 1915.

The outbreak of the war did not bring to an end Elliot Smith's and Rivers' work on the diffusion of cultures; if anything, it gave it extra urgency. For the next few months, their battle was against the opponents of their theory, not the Germans.

As we have seen, when Elliot Smith publicly advanced his thesis that the technique of mummification found in the Torres Straits must

have been brought from Egypt he had been attacked by both Alfred
Haddon and the Oxford archaeologist J. N. L. Myres. Smarting from
these blows, Elliot Smith returned to Manchester late in 1914 and made
contact with William James Perry, a former student of Rivers who
was now working in Yorkshire as a schoolmaster. Then twenty-seven
years old, Perry had come up to Selwyn College, Cambridge, to read
mathematics but had developed an interest in anthropology after
hearing Haddon and Rivers lecture. On graduating he became a maths
teacher but stayed in touch with Rivers, who encouraged him to
research the prehistory of Indonesia, which had an obvious relevance
to his theories about the peopling of the Pacific islands. Perry had
learnt Dutch in order to read missionaries' accounts of native legends
and written a manuscript of a book, which he now sent to Elliot
Smith. The great man was delighted. The work was 'convincing and
conclusive' he wrote on 1 February 1915. Perry had 'unravelled the
apparently hopeless tangle' of Indonesian history and shown that
'there had been an immigration into Indonesia (from the west) of a
people who introduced megalithic ideas, sun-worship and phallism,
and many other distinctive practices and traditions'. He had therefore
provided Elliot Smith with 'all the links in the cultural chain' between
Egypt and Melanesia that he needed. The two men began to correspond feverishly: soon Elliot Smith's wife worried that Perry might
be 'dissipating his marriage portion in postage stamps'. For the rest
of his life, Perry would be Elliot Smith's factotum, busily scouring
libraries for facts with which to support his master's latest theory,
subordinating his pleasant but colourless personality to Elliot Smith's
powerful one.[7]

Armed with Perry's new material, Elliot Smith fought back, accusing
Myres of 'casuistry not worthy of a medieval theologian'. He was
going to 'drop a 15-inch shell into the camp of the "proper" people',
he told Alfred Haddon, and 'use the rest of my ammunition to pick
off any heads that expose themselves after the explosion'. There
remained, though, one major problem with the diffusion thesis – Elliot
Smith had to provide some motive to explain the migrations of culture
which he was postulating; a reason *why* men should have embarked
on great voyages across the oceans. But here, too, a solution appeared,
as if by magic.[8]

In March 1915 Elliot Smith sent Perry a map showing the

distribution around the world of areas in which the presence of megalithic cultures had been demonstrated, telling him also that an American scholar had recently claimed that the 'founders of American civilisation had shown a particular interest in pearls and precious metals'. According to Rivers' later account, 'a chance examination of an economic atlas enabled Perry to combine these two items of information, for he saw that the distribution of pearl shells in the Pacific Ocean agreed so nearly with Elliot Smith's megalithic areas that the presence of pearls would provide sufficient motive for the settlements in these regions. Further examination showed that, in inland areas, the chief motive of the megalithic settlements was supplied by the presence of gold and other forms of wealth.' Elliot Smith was delighted. 'I have no doubt whatsoever that the Phoenicians were the distributors of the "goods" we are dealing with', he wrote to Perry. 'It was loot that led them to colonise certain definite spots . . . How far we have progressed in the last six weeks.'[9]

Elliot Smith was busy running his anatomy department and involved in another controversy to which we will come later – so he relied on Perry's research to keep things moving. As a major figure at the University of Manchester, Elliot Smith had a captive audience for anything he said and a willing publisher for whatever he wrote. The anthropologist Daryll Forde remembered him saying that his work in cultural anthropology 'was, as it were, the lighter side of his life . . . these were, so to speak, recreational activities when he was not having to teach anatomy in the university'. However, when William Rivers returned from the southern hemisphere the tone of Elliot Smith's letters to Perry became noticeably more sober. There was a definite sense of having to account to the headmaster. Rivers visited Perry to tell him of the 'pitfalls' he should avoid and cross-examined Elliot Smith. 'I put my point of view before Rivers and although he saw its reasonableness he wants the evidence in full', Elliot Smith wrote to Perry in July 1915. 'This of course would act more potently if we give it to him in one dose!'[10]

Some of Elliot Smith's and Perry's methodology was amateurish and their evidence contrived. They were particularly vague about chronology; for them, as Daryll Forde later remarked, 'the non-literate populations . . . were timeless snapshots, and these could be taken as stills, so to speak, which had no historical depth behind them'. When they presented new evidence for their thesis to the British Association meeting in Manchester in September 1915 they ran into stern opposition. Few in the audience were convinced that the Phoenicians had been the chief agents in distributing Egyptian civilisation across the world. Sir Arthur Evans, the excavator of Knossos, was dismissive, the Egyptologist Flinders Petrie 'emphasised the need for greater precision in dating the facts with which they dealt', and Sir Richard Temple, a respected former colonial administrator, was unhappy about the use of Indian evidence. Even Rivers' friend, Charles Seligman, questioned the basic methodology, by expressing doubts whether unusual customs or devices found in far-flung places could be considered proof of common influence.[11]

It was left to Rivers to come to his friends' rescue. He did not fail them. Rivers began by admitting to the meeting that, although he had for some time believed in diffusion, he had not initially been convinced by Elliot Smith's attempts to prove that all megalithic monuments were derived from the Egyptian *mastaba* or tomb – partly because he was prejudiced against all theories which saw Egypt as the source of culture; partly because he naturally distrusted simple explanations. What, then, had made him change his mind? For him, the Torres Straits mummy was the decisive factor. 'If we accept the independent origin of mummification in Torres Straits,' Rivers argued, 'we are forced to believe that, in a climate most unsuited for such experiments, the rude savages of these islands invented a procedure which took the highly civilised Egyptian many centuries of patient research to attain.' Rivers preferred not to adopt such a miracle as a scientific explanation:

> In this case there can, of course, be no shadow of doubt that the movement was from Egypt to Torres Straits and not in the reverse direction. I no longer hesitate to believe that the groups of customs and beliefs forming the complex most suitably known as megalithic, developed in Egypt and spread thence to the many parts of the world where we find evidence of its existence at the present time.

However, Rivers did have one caveat. Elliot Smith had argued that the decisive migration had taken place in the eighth century BC: from the

Melanesian evidence, Rivers believed that there had been more than one migration.[12]

By this time, however, the war had begun to claim both Rivers and Elliot Smith.

In July 1915, during the summer vacation, Grafton Elliot Smith was sent by the Medical Research Committee to help at an unusual hospital for shell-shocked soldiers which had been established at Maghull, north of Liverpool. There he found a kindred spirit. Thanks to the upheavals of war, Maghull Hospital was being run by the pathologist at Lancaster Asylum, Dr Richard Rows. Before the war, Rows had campaigned for mental health reform, advocating the establishment in Britain of a system of clinics for helping ordinary people in the early stages of their mental disorder, rather than waiting until they were in such a bad way that they had to be committed to an asylum. Calling for proper training in psychological medicine, he had also begun to experiment clumsily with the new ideas for understanding the mind and treating its afflictions then being developed in Europe. The war now enabled Rows to realise his dream: with the backing of the grandees of the Medical Research Committee and (more warily) of the War Office, he had turned Maghull into a revolutionary establishment, a place where the 'early treatment of mental disorders' by psychological methods could be tried out. Thanks to him, ordinary soldiers were now being treated by very grand doctors.[13]

Elliott Smith was captivated at once. 'I am doing *real* psychology here on soldiers suffering from shock', he wrote. 'The work is extraordinarily interesting and illustrative.' He immediately summoned Rivers, now working at a hospital near Cambridge, to join him. 'Elliot Smith is now full of psychoanalysis and wants me to go over what they are doing', Rivers wrote to Perry; adding, a week later, 'Elliot Smith and I found evident traces of megalithic culture at Maghull.'[14]

Rivers moved to Maghull. But he took some time to get into shell shock work; his mind was still in Melanesia. In the evenings he mystified his colleagues with references to 'stone seats, megaliths, rags hung on trees, dragons, cranial deformation, circumcision, [and] mummification'. But Elliot Smith leapt boldly in, identifying at once with Dr Rows' sense of mission. The hospital was sometimes criticised within the army for ignoring its military duties:

> In the early days . . . inspections by the type of Brass Hat who professed
> to think that all the half-thousand patients were skrimshankers and
> malingerers, though infrequent, did happen. [Rows] had his way with
> these visitors. After giving them a good meal, he saw that they got the
> full treatment: a leisurely tour of mental defectives, schizophrenics and
> maniacs.

Elliot Smith used his medical prestige to shield Dr Rows against the
War Office, and, in a combative 1916 article in the medical journal the
Lancet, compared Maghull's successes with neurasthenic soldiers with
the hostility and lack of interest in such cases shown by ordinary
doctors and 'certain neurologists'. Then in 1917, he and his Manchester
colleague T. H. Pear published *Shell Shock and its Lessons*, a brilliant
polemic which argued that Maghull's wartime work should serve as
a beacon for post-war mental health reform. The war, wrote Smith
and Pear, had shown that *anyone* could develop a psychoneurosis if
he was exposed to a sufficiently difficult environment; and thus
disproved completely pre-war ideas that mental illness was hereditary
and untreatable. Neurologists and asylum doctors responded by
comparing the family and personal histories of one hundred wounded
soldiers and one hundred shell shock cases at the Maudsley Hospital
in South London. They found that seventy-four per cent of the psycho-
neurotics had pre-war 'form'.[15]

Elliot Smith's genius for propaganda glossed over what actually
happened at Maghull. Dr Rows was indeed able to bring to the hospital
a 'brilliant band' of mental health specialists who taught the rudiments
of the new psychology to military doctors being trained to treat shell-
shocked patients – a strange compote of Freud, Jung, Janet and other
foreign prophets. But the most influential part of their curriculum
was the work of the French pioneer of psychotherapy Jules Dejerine,
who advocated simple emotional sympathy as the most effective form
of treatment. And, even then, applying theory at a practical level was
not always easy. Learned polymaths such as Elliot Smith were astonished at the level of ignorance in their patients, such as the soldier
who believed that nocturnal emissions would affect his brain and drive
him mad, as '"trashy" adolescent literature had led him to believe'.
When the structure and functions of the sexual organs were explained
to him in an elementary way, he exclaimed 'Why wasn't I taught this

when I was a lad, for then I would have been spared all this trouble?' It was difficult to practise psychoanalysis with such patients. Besides, the men had their own, simpler, priorities. When, for example, it was discovered that the interpretation of dreams might determine whether a soldier was returned to the Western Front, they began to withhold them from the doctors.[16]

In fact, Maghull's most effective form of therapy had little to do with the theoretical teaching done there. It derived more from Elliot Smith's hatred of authority. Refusing to put on uniform, insisting on maintaining the values of a civilian, he used his prestige to make sure that the hospital ignored the wishes of the military and returned few soldiers to the front line. The army hierarchy, for its part slowly and reluctantly began to accept that it was useless to return men with chronic neurosis to front-line duty. 'A medical general came almost secretly to the hospital,' a doctor recalled later, 'and said would we please discharge as many from the hospital as possible, though we must not say he said so.' In the year to 30 June 1917, only 20.9 per cent of the 731 patients discharged from the hospital returned to military duty, while sixty-five per cent went back to civilian life, twelve per cent went to other hospitals, and one per cent went to civilian mental hospitals.[17]

Rivers meanwhile remained preoccupied by anthropology – preparing an important set of lectures to be given in London the following year – and by the reappearance of John Layard in his life. His young protégé had returned to England in 1915, starved and weakened by his year on Atchin Island. Always emotionally fragile, and struggling now to adjust to a society mobilised for war, he was quite unable to cope when his father, who was living in a rooming house in Tooting after his house in Suffolk was requisitioned by the military, went off his head. In desperation, Layard wrote to Rivers, asking for help.[18]

Rivers must have felt guilty about the way he had abandoned Layard on Atchin, because he immediately came down to London and organised for Layard's father to be sent to a private mental hospital. But this was not the end of Layard's troubles. As he lay in bed in Cambridge, trying unsuccessfully to write up his anthropological notes, he became convinced that he was 'going batty' himself. He appealed once again to Rivers, who invited Layard to Devon where he was travelling for a period of restorative rest, in February 1916. Here he installed Layard, first in a friend's house, and then in a hotel at Salcombe.[19]

By this time Rivers, after much distraction, was finally beginning to engage with the shell-shocked soldiers at Maghull. His interest was not aroused by the human plight of the men – as McDougall's had been – but by an intellectual problem: he became fascinated by the meaning of dreams. By coming to Maghull, Rivers was joining what Elliot Smith called a 'society in which the interpretation of dreams and the discussion of mental conflicts formed the staple subjects of conversation'. Staff at the hospital had soon noticed that 'patients who, when awake, joked, played cards and billiards [and] attended dances', also 'raved and sleepwalked at night, giving the intelligent nurse the opportunity for revealing reports'. One patient was heard to say in his dreams, 'It was an accidental shot, Mao'or, it was not my fault.' Another soldier's dreams 'always began with some terrible experiences in the trenches and then turned to some sexual acts with women, usually with his wife and he woke to find the clothes disturbed and that he had "lost nature"'. The men were very alarmed by it all: 'Questions such as "Why do I get these terrors?" "Why am I always seeing those things that happened in France?" "Why am I so irritable?" "Why do I get so upset?" are frequently heard from patients', Dr Rows recorded. 'In almost every instance the memory of some disturbing past experience will be found acting as the cause.'[20]

Although it had been known for centuries that people who live through terrifying, or emotionally disturbing, experiences dream about them afterwards, war dreams only really began to be systematically studied in the Great War; and by 1916, there was quite a medical literature about them. Doctors noticed that war dreams, unlike peacetime anxiety dreams, were repetitive, the dreamer usually returning night after night to the same scene of horror – although whether it was the faithful repetition of an actual experience was a matter of argument.

Rivers recognised at once the importance of soldiers' dreams to a doctor treating the war neuroses. But he went about the problem in his elaborate, intellectual way. On arriving at Maghull he began to read Freud's recently translated book on the interpretation of dreams. At first it disappointed him: Rivers found Freud's reading of his own highly complicated dreams 'forced and arbitrary' and his general method so 'unscientific that it might be used to prove anything'. So, initially, Rivers failed to engage with the subject. Partly, he was still engrossed in ethnology; but he also found the ordinary soldiers at Maghull uninteresting as patients. For all his skill with tribesmen, he did not establish much rapport with ordinary Tommies and later complained that while there he had 'little opportunity for testing dream-interpretation practically'; such of the men's dreams as he was able to get hold of were usually very straightforward and confirmed Freud's theory that dreams express the fulfilment of an unconscious wish. The patients dreamt that they were back in bed with their wives or that they were sent back to France, only for peace to be declared. 'The dreams of uneducated persons are exceedingly simple and their meaning is often transparent', Rivers later wrote. 'It was only when I began work in Scotland that my growing interest in the psychological problems suggested by war-neurosis began to compete and conflict with my interest in ethnology.'[21]

The change of attitude was brought about by his transfer, at the end of 1916, to Craiglockhart, a hospital for officers based in a 'decayed hydro' near Edinburgh; and by a dream he had soon after arriving there. The dream had nothing to do with the war or his patients, but expressed the professional doubts Rivers had recently felt. There was now a chance that he might become president of the Royal Anthropological Society – following in the footsteps of his uncle, James Hunt – and, while one part of him eagerly craved this honour, the other dreaded the travel and committee work it would involve.[22]

The dream seemed therefore to confirm two of Freud's most important assertions: that dreams are about wish-fulfilment and that their manifest content is often a transformation of the latent wishes of the sleeper, which appear in a disguised form. This 'presidency' dream, Rivers recalled, 'went far to convince me of the main lines of the Freudian position', and from now on he accepted the Freudian method of interpreting dreams. He immediately reread *The Interpretation of Dreams* and in March 1917 gave a paper to the Edinburgh Pathological Club, arguing that there was a risk of an important truth being lost in the current very polarised medical debate about Freud: namely, that Freud's theory of the unconscious was of far wider application than recent medical critics allowed. In fact, Rivers argued, the idea of the unconscious was already present in the psychological theory of instincts; similarly, the idea that mental conflict played a role in neurosis had been around for a long time. What made Freud's work so noteworthy, however, was his scheme of the nature of the opponents in the conflict, discussing the role of forgetting as an example. Freud was, in Rivers' view, not unique in using the idea that the mind dissociated the memory of horrible experiences, but his great contribution was in providing an *explanation* of forgetting. The war had provided a unique opportunity to test this theory. While it had shown conclusively that Freud was wrong to insist that all neuroses were sexual in origin it had also confirmed Freud's theory of forgetting. It was clear, therefore, that, in treating the war neuroses, repression was not the answer.[23]

Rivers' paper had the tone of brisk self-confidence. Indeed, he was already applying his own theories, using dreams as the key to understanding the conflicts in his patients' minds. The issue which preoccupied Rivers at Craiglockhart was – what advice he should give to his patients about their horrible war experiences. Should they think about them or try to forget them? Most had been firmly told by friends, relatives or doctors to forget them ('Put it out of your mind, old boy'), to 'lead their thoughts to other topics, to beautiful scenery and to other pleasant aspects of experience'. But Rivers realised it was not that easy. One of his patients had been buried by a shell explosion in France and, after remaining on duty for several more months, had then collapsed after 'a very terrifying experience in which he had gone out to seek a fellow officer and found his body blown to pieces with head and limbs lying separated from his trunk'.

> From that time he had been haunted by night by the vision of his dead and mutilated friend. When he slept he had nightmares in which his friend appeared, sometimes as he had seen him mangled in the field, sometimes in the still more terrifying aspect of one whose limbs and features had been eaten away by leprosy. The mutilated or leprous officer of the dream would come nearer and nearer until the patient suddenly awoke pouring with sweat and in a state of utmost terror.

This man had tried to suppress his memories by day, 'only to bring them upon him with redoubled force and horror when slept'. In treating him, Rivers realised, he had to find some way of letting the patient confront his memories that would also reduce the horror. The solution was to suggest to him that the friend's mangled body proved that he had been killed outright and thus spared the drawn-out agony of dying by wounds. Whereupon, Rivers said, the patient 'brightened at once' and agreed to try to think of the experience in a more positive light. For several nights the dream did not recur. Then, one night:

> He dreamt that he went out into no-man's-land to seek his friend, saw his mangled body just as in his other dreams, but without the horror that had always been present. He knelt beside his friend to save for the relatives any objects of value, a pious duty he had fulfilled in the actual scene, and as he was taking off the Sam Browne belt he woke with none of the horror and terror of the past, but weeping gently feeling only grief for the loss of a friend.

This wasn't quite the end of it: the horrifying dream did recur occasionally, but in a few months the man had left hospital and was regaining his normal health and strength.

Rivers was at first surprised by the change both in the subject matter and the emotional character of the dream, but with experience he came to recognise it as one of the signs that a patient was improving. Realistic scenes of war, repeated in identical form night after night, would give way to other images, such as terrifying animals, and the accompanying emotion of fear would start to die down.

But there were limits to what could be achieved by the simple lifting of repressed war memories. Sometimes a patient's experience was so horrible that Rivers could not offer him a positive way to confront it. There was a young officer

> who was flung down by the explosion of a shell so that his face struck the distended abdomen of a German several days dead, the impact of the fall rupturing the swollen corpse. Before he lost consciousness, the patient had clearly realised his situation, and knew that the substance that filled his mouth and produced the most horrible sensations of taste and smell was derived from the decomposed entrails of an enemy.

By the time he reached Craiglockhart, this patient was 'striving by every means in his power to keep the disgusting and painful memory from his mind' and Rivers thought it best that he leave the army and go off to the country where he had previously found peace of mind.[24]

Just as Rivers was getting to grips with shell shock he was confronted with the latest stage in the Layard saga. John Layard had insisted on following Rivers to Edinburgh, dragging his family with him, so that he could receive analysis. They found lodgings near Craiglockhart, and, according to Layard, 'Rivers came every morning, stayed an hour and then went away. The whole of life was waiting for Rivers to come and when he had gone away, waiting for him to come the next day.'

We only have Layard's account of what now followed. According to his autobiography, he never found the courage during these daily visits to say the 'most important thing'; that is, the 'transference and the love' he felt for Rivers. Then one day, the situation came to a head: 'I said, "I love you but you're sending me off my head. But I can't see you any more." I remember Rivers almost trembling and blanching when I said that. He went and that was the last time I ever saw him.' As Layard wrote later, 'Rivers had obviously not recognised the whole homosexual content of our relationship, probably [on] both sides.'[25]*

Layard's departure seemed finally to liberate Rivers. He was now fully engaged with his war work and the acknowledged master among the staff at Craiglockhart. 'When a fresh convoy of patients arrived,' a colleague later recalled, 'Captain Rivers walked round them and took his pick. Nearly all the interesting patients floated his way.' In an article on psychotherapy he wrote during the war, Rivers spoke of the importance of the 'personality of the healer'. Indeed many elements came together in his therapeutic persona: rigorous clinical training at Queen Square and Bethlem; research on fatigue; experience, as an overworked house physician, of pressures similar to those faced by young officers in the war; years patiently questioning the

---

* Layard's later life was unhappy and he never really fulfilled his potential as an anthropologist. In the 1920s he was treated with some success by Homer Lane, a maverick American psychologist. He next turned up in Berlin, taking a prominent part in its thriving expatriate homosexual scene. Back in London, Layard met Jung and was rescued from insanity by Doris Dingwall, an anatomist at University College London, whom he married in 1943. The same year he finally published *Stone Men of Malekula*. After the war, Layard practised as a psychotherapist. My mother was one of his patients. He died in 1974. 'His unique, if somewhat eccentric, contribution to British anthropology has not won the attention it deserves.' (J. MacClancy, *ODNB*.)

tribesmen of Melanesia. As Elliot Smith remarked, 'the measures taken to discover the causes of the soldiers' mental disabilities were so similar to those he had been using in Melanesia into the social and magico-religious problems of peoples of lowly culture'. Henry Head spoke later of 'his remarkable power of gaining the confidence of young men' and his 'vivid interest in the personality of each individual under his care'. More educated patients were astonished to find someone like him working as a military doctor. 'I met one rather interesting man up here,' the writer Max Plowman wrote to his wife on 29 April 1917, 'A Dr – who's a Professor of Psychology at Cambridge. I was talking to him about Freud's book on dreams and he lent me Hart's *Psychology of Insanity* as an introduction to it.' It is thanks to one of these patients that we have a remarkable record of Rivers at this time.[26]

When Siegfried Sassoon arrived at Craiglockhart in August 1917, he was thirty and some way along the road from agreeable Georgian versifier to embittered poet of war. Far from being the harmless countryman hunting in the woods and playing cricket on the village greens of Kent that he presented in his autobiographies, Sassoon was an exceptional and complicated person. His father, who came from a rich and powerful Jewish family, had married outside the faith, and then left home when Siegfried was seven. As a result, the dominating influence of his life was his mother, who belonged to an artistic dynasty: one of his uncles, Sir Hamo Thornycroft, was Britain's most successful sculptor; his aunt Rachel Beer edited both the *Sunday Times* and the *Observer*; and his cousin Philip Sassoon was private secretary to General Haig.[27]

Family wealth had enabled Sassoon to leave Cambridge without a degree and to live as a country gentleman until the war came. Commissioned into the Royal Welch Fusiliers, his exploits on nightly trench-raiding parties soon gained him a reputation for bravery, the nickname Mad Jack, and the Military Cross; but his military career was complicated by his homosexuality, of which he had been aware since Cambridge, and by the powerful feelings of love and admiration which he felt for the men he commanded.

In the autumn of 1917, while back in England recovering from wounds, and much influenced by pacifists such as Bertrand Russell, Sassoon wrote a declaration of opposition to the war which he somehow managed to get read out in the House of Commons. He expected to be court-martialled but instead – after deft interventions by his friends – the authorities declared that he was suffering from shell shock and had him sent to Craiglockhart. There he became Rivers' charge.

Sassoon's initial responses to Craiglockhart were those of a man of his class. 'I hope you aren't worried about my social position', he wrote to Lady Ottoline Morrell. 'My fellow officers are 160 more-or-less dotty officers. A great many of them are degenerate-looking. A few are genuine cases of shell shock etc.' He loathed the 'truly awful atmosphere of this place of washouts and shattered nerves'. Rivers, though, was 'a sensible man who doesn't say anything silly. But his arguments don't make any impression on me. He doesn't pretend that my nerves are wrong, but regards my attitude as normal.' Rivers, for his part, wrote in his admission notes on Sassoon: 'The patient is a healthy-looking man of good physique.'

> There are no physical signs of any disorder of the nervous system. He discusses his recent actions and their motives in a perfectly intelligent and rational way, and there is no evidence of any excitement or depression. He recognises that his view of warfare is tinged by his feeling about the death of friends and of the men who were under his command in France.[28]

Doctor and patient soon struck up a close relationship, the great bond being that they came from the same part of Kent: Sassoon had grown up at Paddock Wood, a few miles from Tonbridge. Their common homosexuality may also have been an unspoken affinity, but, as Rivers had shown in his dealings with John Layard, he was not a man to externalise his feelings. The relationship with Sassoon was more that of father and son.*

In one of their first 'friendly confabulations', Sassoon asked Rivers if he was suffering from shell shock:

---

* Countless writers, from Paul Fussell to Adam Kuper to the Tonbridge School website, have stated that Wilfred Owen was also a patient of Rivers' at Craiglockhart. In fact, Owen was treated by Rivers' colleague, Arthur Brock, an eccentric Scotsman who believed in ergo-therapy, the cure by work, and showed no interest in dreams. This was fortunate for Owen: it meant that, instead of telling his dreams to his therapist, he wrote poems about them. See Dominic Hibberd, *Wilfred Owen. The Last Year 1917–1918* (1992).

'Certainly not,' he replied.

'What have I got then?'

'Well, you appear to be suffering from an anti-war complex.' We both of us laughed at that.

In his later memoir, Sassoon described how, three evenings a week, he went along to Rivers' room 'to give [his] anti-war complex an airing'. They talked mainly about what European politicians and were saying about the war; both were readers of the *Cambridge Magazine* which summarised the foreign press. Sassoon's first war poems, satirising civilians, non-combatants and the old men who were sending the young to their deaths, had begun to appear in it. Rivers was able to help Sassoon write at Craiglockhart by gently removing an over-talkative room-mate.[29]

The relationship between the two men was further complicated by the war and Rivers' role in the Royal Army Medical Corps. By the autumn of 1917 the young poet was having dreams in which the soldiers of his platoon reproached him for abandoning them in France. He was increasingly restless at Craiglockhart and Rivers eventually persuaded him that his best course was to return to the war. When Sassoon missed a medical board, Rivers deftly handled the bureaucratic fallout.

But Rivers himself was approaching a crisis. His growing discomfort with his duty as an officer in the Royal Army Medical Corps – which required him to send young men back into a war he himself had increasing doubts about – was reflected in his dreams, and the unremitting demands of his work at Craiglockhart was affecting his fragile health, forcing him to take repeated breaks. Sassoon later recalled Rivers 'sitting at his table in the late summer twilight, with his spectacles pushed up on his forehead and his hands clasped on one knee; always communicating his integrity of mind; never revealing that he was weary as he must have been after long days of exceptionally tiring work on those war neuroses which demanded such an exercise of sympathy and detachment combined'. But Rivers' years as a house physician at Queen Square in the 1890s had taught him to heed the warning signs.[30]

In October 1917, Rivers was offered the chance to work in London, at an Air Force hospital, alongside his friend Henry Head. This was a hugely attractive proposition, not least because in the capital it would be possible to combine psychiatry with his ethnographic work. On the other hand, it would mean abandoning his patients and leaving in the lurch Dr Bryce, the medical superintendent at Craiglockhart, who had been hugely supportive of Rivers and was constantly battling with the War Office. In the end, Rivers decided to leave. He could not risk another collapse such as had been narrowly avoided at Queen Square in the 1890s. No one at Craiglockhart held it against him.

In London, Rivers could join in the medical debate on shell shock. Earlier in the year, his paper arguing that Freud's work could be useful in treating the war neuroses had provoked a strong response. Rivers had many letters, some commending him for bravely advocating the ideas of an enemy in wartime, most congratulating him for denying the sexual origin of neurosis. 'I am very glad that you look askance at the pornographic side of the work of Freud and his followers and that you point out the value of his method apart from that', Dr R. Percy Smith, his old boss at Bethlem Hospital, had written. 'The psychoanalyst as a rule seems to plumb for the sexual aside and I am sure often suggests it to the patient's mind and I have seen this done with disastrous results.' Dr Smith was in the chair in November 1917 when Rivers gave a paper on 'The Repression of War Experience' to the Royal Society of Medicine, and in the audience were several leading psychoanalytic figures such as Ernest Jones and the Jungians David Eder and Maurice Nicoll, who hastened to congratulate Rivers. In the paper Rivers referred in coded terms to his own repression of his own sexuality. 'Repression', he declared, 'is not itself a pathological process nor is it necessarily the cause of pathological states. On the contrary, it is a necessary element in education and all social progress.' The problem in wartime, however, was that the training in repression normally spread over years had had to be telescoped into short spaces of time.'[31]

In London, Rivers was able to go up a professional gear. He lived in Hampstead near the Heads, and worked alongside Henry Head in the Air Force Hospital at Mount Vernon and at the Empire Hospital in Victoria, where his friend was studying the effects of brain injury. With his usual thoroughness, Rivers steeped himself in the problems of aviation medicine. He took numerous flights and made the pilots 'loop the loop' and carry out stunts and manoeuvres in the air, so that he had enough experience of flying to be able to treat his patients and to test candidates satisfactorily. The clinical load was less punishing than at Craiglockhart: Rivers found 'the usual cheerful, irresponsible kind of youth with whom one was accustomed to deal in the Air Force, very different from the man weighed down by responsibilities and anxieties with whom I had been accustomed to deal in the army'.[32]

Though Rivers had had to leave Craiglockhart, his successful work there gave him much greater confidence. Grateful former patients came and saw him. Friends noticed that he was now more worldly and outgoing. He began to be involved in committees and advisory work. 'I am having a splendid time here, getting some decent writing done', he wrote to Sassoon in February 1918. 'My only trouble is that I am getting too many irons in the fire. "Affairs" are so seductive and they so disturb the contemplative life in which I do my best work.'[33]

Rivers' wider experience in London led him to look again at the nervous disorders of the war from what he called 'a general biological standpoint'. Although he had advocated using Freud's theories as a therapeutic tool, when it came to advancing an overall theory of shell shock, Rivers went back to his neurological grounding under John Hughlings Jackson at Queen Square in the 1890s and to his work with Henry Head in the 1900s which had distinguished between 'protopathic' and 'epicritic' levels in the nervous system; he also drew on the London surgeon Wilfred Trotter's book *The Instinct of the Herd* and borrowed (without acknowledgement) from William McDougall's work on instincts. It was, however, noticeable that Rivers, unlike many of his contemporaries, made no attempt to apply Walter Bradford Cannon's work on the role of the endocrine system in the body's response to fear to his explanation of shell shock.

Psychoneuroses, Rivers now argued, take place when the balance between the older, more primitive instincts (such as sex and self-preservation) and the newer controlling 'forces' such as duty, fidelity, intelligence and religion is disturbed by shock, strain, illness or fatigue. In civilian life it is usually the sexual instincts which are thus exposed; in war, it is those of self-preservation. Psychoneuroses, Rivers suggested, can be seen as attempts 'to restore the balance between instinctive and controlling forces'.

Hysteria – which in wartime was usually found in less educated soldiers – was a way of escaping from this conflict rather than facing it, and normally the result of letting go of 'the modifying principle based on intelligence'. In hysteria, said Rivers, an 'ancient instinctive reaction to danger' – namely immobility – was reactivated, causing paralysis, dumbness, and insensitivity to stimuli. At the same time, hysteria produced an extraordinary suggestibility, particularly in soldiers put through the ritual of military training. Hence the ease with which hysterics could be hypnotised or would start copying each other's symptoms.[34]

Most of Rivers' patients, however, had been officers, suffering not from hysteria but from 'anxiety neuroses'. These, too, were a form of regression, but of a less profound kind. These men did not go back to an earlier evolutionary level, but to 'the character of infancy'. Rivers saw the loss of emotional self-control as a return to the violent emotions of childhood. The war dreams of these patients were the same as childish night-terrors and were full of the same imagery, such as terrifying animals. There was the same failure to draw a line between imagination and external reality; in some patients this led to schizophrenia.[35]

In London, Rivers took up ethnography again. In the later years of the war, Grafton Elliot Smith and Will Perry had let their imaginations loose in pursuit of diffusion, roaming freely across art history, mythology, architecture and archaeology, an intellectual voyage charted by Elliot Smith's letters to Perry. After 1915, he had enlarged his area of interest to take in not just the spread of culture from Egypt to the eastern Mediterranean, but the diffusion of Egyptian elements across the Indian Ocean to Asia and even to the Americas; for Elliot Smith such elements as the representation of the elephant in pre-Columbian America, or winged-disc motifs in Mexico and Central America, provided evidence of the Egyptian origin of certain elements of pre-Columbian civilisation. American specialists did not take kindly to this incursion on to their territory and struck back forcefully when Elliot Smith published a 'crude sketch' of his views in August 1916. Why, if there had been Egyptian influences, was there no such thing as a wheeled vehicle in all pre-Columbian America? An invention as useful as that could hardly have fallen into disuse if it had arrived from Egypt. Why was the most elaborate boat to be found in South America a raft of light wood?[36]

Undeterred, Elliot Smith ploughed on. He was, as the historian Adam Stout has written, 'much preoccupied with trying to understand why the Egyptians had gone to such trouble to embalm their leaders, and slowly developed the idea that they travelled far and wide in search of "givers of life" that would restore life to the dead: expeditions that eventually resulted in the colonisation of the world by an elite warrior caste'. This process was not necessarily seen as benign. Indeed both Elliot Smith and Perry were infected by the pessimism and pacifism of wartime. Perry began to argue that warfare was not a natural condition of mankind but the by-product of a sun-worshipping aristocracy which had enslaved peaceful populations around the world. Warfare, Perry declared in a talk in Manchester in February 1918, 'is the means whereby the members of a parasitic ruling class of alien origin endeavour, while exploiting their own subjects, to dominate those surrounding peoples who produce wealth . . . This process of exploitation and domination of the many by the few will last until the common people of the earth recognise their condition and become aware of their power . . . The very patience with which the peoples of the earth have submitted to domination,' Perry concluded, 'and their resignation under the most unjust and cruel treatment, constitute powerful evidence of the innate peacefulness of mankind.' Four months after the October Revolution in Russia, Perry had caught the incendiary smell of political upheaval in the air.[37]

It is not clear how far Rivers went along with all this. There was certainly a diffusionist tinge to his wartime articles on healers and primitive dreams but he also advised Perry not to get too close to Elliot Smith, whom he now described as 'reckless'. He proved 'quite unsympathetic' to Elliot Smith's views on the connection between the origin of cultivation and moon worship, arguing that 'agriculture may have developed independently in many places', and restrained his friend when the latter wanted to resign from the Anthropological Institute and start his own independent journal. But the two men remained close, and for both of them conversations about diffusion provided an escape from the pressures of wartime.[38]

If Rivers' faith in the idea of diffusion did ever falter, it was powerfully reinforced in August 1918. Just as the Allies' final offensive on the Western Front was getting under way, Rivers received a letter from a certain Rev. Charles Elliot Fox, a missionary based on San Cristoval in the Solomon Islands, just to the east of Guadalcanal, which provided the 'first account of the burial customs of the chiefly clan and their worship of the sun and the serpent'. Rivers was so excited that he took leave from his hospital and asked Elliot Smith to join him in the Lake District to devote a week to the undisturbed discussion of the new evidence. For him, this became what the Torres Straits mummy had been for Elliot Smith. 'The resemblance between the mortuary customs of ancient Egypt and modern San Cristoval, so close and extending to so many points of details,' he stated in a lecture the following year, 'makes it incredible that they should have arisen independently in these two regions. We can be confident that mariners imbued with the culture of ancient Egypt, if they were not themselves Egyptians, reached the Solomon Islands in their search for wealth, and that their funeral rites so impressed the people of these far-distant isles that they have persisted to this day.'[39]

Thus encouraged in their thinking, as the war came to an end, Rivers and Elliot Smith began to make bold plans for the future.

PART SIX

'THE WORLD IS DIFFERENT NOW'

Projects and Politics

Elliot Smith, Rivers, Myers and McDougall all came back from the war fired up with a new sense of social engagement, determined to change things. Above all, they wanted to apply in civilian life the insights into human behaviour which they believed the war to have given them. This final, socially minded phase in their careers would take them into new directions.

The end of the First World War found Europe in turmoil. Russia was now ruled by the Bolsheviks and revolutions were breaking out all over Eastern Europe. The old order seemed to be tottering. This climate – combined with the influenza epidemic which swept the continent – produced a strange, febrile mood among British intellectuals, one manifestation of which was a craze for Freud. 'The psychoanalytic ferment here is remarkable', Ernest Jones wrote to Freud from London in January 1919. Everyone talked about the unconscious. 'Even pickpockets are now appealing to their judges to regard their cases from the psychological point of view', one elderly medical man complained. 'It is the fashion at present to analyse everything.'[1]

Charles Myers was strongly affected by the mood of reform. After returning from France late in 1917, he had operated as an intermediary between the military authorities and the now considerable network of shell shock hospitals dotted around the country, many of them staffed by doctors trained at Maghull. The doctors he spoke to – 'a mixed bag of psychiatrists, psychotherapists, psychologists and "pure" neurologists' – had found themselves treating not simply the hysteria and anxiety neuroses caused by the war but also, in many cases, the underlying psychological weaknesses of their patients. This de facto examination of the mental health of the British people left them horrified at the problems which 'normal' people seemed to have. Myers now became a spokesman for this group.[2]

In April 1918 he gave two lectures on 'Present-day Applications of Psychology, with special reference to Industry, Education and Nervous Breakdown', in which he sketched out a vision of a new and much broader kind of psychology than that taught at Cambridge before the war. He noted how experimental psychology had recently realised 'the enormous importance of the study of feeling – alike in observation, memory, thought, decision and action'. Myers compared the conventional laboratory approach to the problems of memory and forgetting to those revealed by the 'new psychology':

> The factor of feeling was expressly eliminated from our experiments on memory. The material learnt in the laboratory consisted commonly of meaningless numbers or of senseless syllables, so that all interest or other references to past experience might be eliminated as far as possible. But now we begin to realise that what is learnt may *never* be forgotten . . . In the cases of nervous breakdown, which have resulted in this war, it is astonishing how early emotional experiences may become revived (perhaps in some distorted form) and become responsible for protracting the emotional condition of the patient.

The high priest of Cambridge psychology seemed to admit that the war provided a better laboratory than his own experiments.[3]

The lesson of the war was clear – a new kind of psychology was needed to take the discipline into such areas as mental abnormality, child development and industrial neurosis. But for this to happen, the profession needed to become less exclusive, change the way the subject was taught, break down the barriers between sanity and madness, and find ways to apply its insights in industry. 'A new class of medical man, educated in the psychological theories and practice which I have described is being trained', Myers declared. 'One centre of instruction [i.e. Maghull] has already sprung into existence during the war and others must be instituted.'[4]

One part of Myers' programme proved easy to achieve. At the end of 1918 he proposed that membership of the British Psychological Society should no longer be confined to a small elite of academics but be available to anyone working in medicine, education, industry or the army, and interested in the subject. This was immediately accepted; the society was revitalised.[5] Other elements proved more difficult. Both Myers and William Rivers wanted to place Cambridge itself at the vanguard of 'new psychology'. At the same time, they argued, a new kind of medical institution needed to be created, the mental health clinic: the sort of halfway house between the community and the asylum, offering treatment to those suffering from the early stages of mental illness, that Ronald Rows had been calling for in 1910 and for which Maghull Hospital now provided a wartime model. By establishing such a clinic, with Colonel Rows, the maestro of Maghull, at its head, Cambridge would show the rest of Britain the way forward.

For a while, Rivers' eloquence and authority and Myers' organising skills gained some headway for these ideas. But then a reaction set in. The local authority was fearful that the clinic would require new legislation and incur extra costs; the university, taking its cue from Sir Clifford Allbutt, the 82-year-old Regius Professor of Physic, turned against it; Allbutt had been alienated by the 'wild rise of psycho-analysis'. Worst of all, public support for the initiative was not forth-coming. The effete denizens of Bloomsbury might clamour for Freud, but out in the sober Fens there was less demand.[6]

Similarly, the scheme to revamp the psychology department was defeated. The university decided that in Cambridge psychologists must stick to experimental work – 'psychology with instruments' – and not get involved in new-fangled fashionable ideas. Myers was promoted from lecturer to reader, but – at the insistence of James Ward, the professor of philosophy – in 'experimental psychology'. For Myers, this was the last straw. His wartime experiences had changed him. 'He was not the same. He was never to be the same again', Frederic Bartlett found. 'The radiant smile was seen less frequently, he tired more readily. Much of his natural buoyancy and liveliness had gone. Perhaps most noticeable of all, his sensitivity to opposition and criti-cism had become far more marked.' Early in 1922, Myers resigned his university post, turned his back on the laboratory he had paid for with his own money, and went off to London to pursue his idea of applying psychology in industry.[7]

For Rivers the campaign to reform Cambridge psychology was just one facet of a complex post-war readjustment. He had resigned his university position in 1916 but in December 1918 St John's College appointed him to the newly created post of praelector in natural science – which was designed to provide more contact between undergraduates working in a subject and senior research workers in that field, 'a system of mentoring *avant la lettre*', as the college history puts it. When his friend Lewis Shore asked Rivers if he would take the job, 'his eyes shone with a new light that I had not seen before, and he paced his rooms several minutes full of delight'. Rivers was by general consensus a great success in his new, ill-defined role. 'Rivers is hard at work, and there is already a new spirit about the College', a supporter wrote in January 1919. But, this being Oxbridge, there were factions: T. R. Glover, a classical scholar and diarist, found Rivers 'a bit doctrinaire and dogmatic'. The students adored him. Rivers' room in New Court seemed to his new friend, the novelist Arnold Bennett, 'like a market square':

> Undergraduates came into it at nearly all hours to discuss the intellectual news of the day. They came for breakfast, but I think from ten to one he would not have them. During those hours he used his typewriter.
>
> His manner to young seekers after wisdom, and to young men who were prepared to teach him a thing or two, was divine. I have sat aside on the sofa and listened to hundreds of these interviews. They were touching in the eager crudity of the visitors, the mature, suave, broad-sweeping sagacity and experience of the Director of Studies, and the fallacious but charming equality which the elder established and maintained between the two.[8]

Rivers too had been transformed by the war. Everyone noticed how much more confident, relaxed and sociable he had become. 'When he returned once more to College', Shore recalled, 'a certain reticence, a certain shyness, and a certain difficulty of approach which some of us had always felt with him had disappeared.' According to Charles Myers, 'diffidence gave way to confidence, hesitation to certainty, reticence to outspokenness, a somewhat laboured literary style to one remarkable for its ease and charm'. Where formerly Rivers had shunned company and avoided public duties now he freely sought out both. 'The world's different now', he told Frederic Bartlett. 'Things are in a mess, and no matter how big a bother it is we've simply got to help put them straight.'⁹

What had brought about this remarkable change? Recommending Rivers to his American publisher, Arnold Bennett wrote that 'He has accomplished a number of very wonderful cures in cases of mental derangement among soldiers.' Certainly Rivers' success as a psychotherapist had given him a new sense of fulfilment and self-worth, and made his name known among a broader circle. But Frederic Bartlett offered a different explanation: 'perhaps the war helped him to find out that English people – particularly young ones – were nearly as interesting as Melanesians'. He also thought that Rivers, having got his great work on Melanesian society off his chest, 'could let all kinds of tendencies and capacities that he had held under and brought into submission have their full swing . . . The change was not essentially in the man half so much as in the place of his work.' It is equally likely that the prolonged self-analysis which Rivers underwent during the war, and which he had charted in his book on dreams, gave him new confidence.

Rivers had also had a good war, intellectually. His side had won. When *Instinct and Unconscious*, which brought together his wartime writing on shell shock, came out in 1921, the eminent Queen Square neurologist Samuel Kinnear Wilson conceded that both psychiatrists and neurologists had been found wanting in the war. 'It [soon] became obvious that only psychological investigation would satisfy the searcher after cause and effect and many who had been but little attracted by the Freudian theory found themselves applying it to the human problems before them. In the sequel . . . we have left our insular conservatism behind, and there has arisen a school of clinical psychologists whose contributions to the theory and practice of psychological medicine have largely made up for the barrenness of earlier years.'¹⁰

Rivers was the leader of that school. So where would he lead his followers? In fact the head of steam, the clamour for a new kind of psychological medicine, soon subsided. The failure to create a psychological clinic or to reform the psychology department in Cambridge played a large part, but Rivers also shifted his position. Having acted as an evangelist for Freud in the two important articles he published in 1917, he found himself courted by the English Freudians and agreed to hold several positions in the newly created British Psychoanalytical

Society. At the same time the recently enlarged British Psychological Society shared platforms with the psychoanalysts.[11]

Gradually, however, relations cooled. Rivers ultimately came up with a 'biological' theory of shell shock – as a reversion to earlier, more primitive, levels in the nervous system – which drew more on the British neurological tradition than on Freud. But he also reacted against the popular craze for psychoanlaysis. In 1921 the magazine *Discovery* reported that 'Thanks largely to its successful application to shell shock cases', psychoanalysis 'took the public by storm towards the end of the war'. That was enough to put Rivers off. And while wartime clinical experience had shown the usefulness of Freudian techniques for some patients, it had also brought out their limitations: psychoanalysis was too time-consuming and quite useless with half-educated soldiers. What soon emerged in Britain was an eclectic tradition which accepted Freud's theory of the dynamic unconscious, and much of his account of mental conflict, but rejected his determinism and emphasis on sex, while freely borrowing from Janet, Jung and Alfred Adler and, above all, respecting existing moral and religious beliefs. Rivers was very much in this 'sane', 'sensible' and 'ethical' stream. In November 1921 he told Ernest Jones, 'I am becoming progressively more and more doubtful about the value of psycho-analysis as at present conducted, especially on the ground of its produc-tion of undue dependence and loss of critical faculty, and consequently have a great dislike to undertaking any kind of responsibility in connec-tion with it.' Replying, Jones regretted that where once the two men had engaged in mutual criticism in a spirit of scientific cooperation, 'now it seems that I must draw another conclusion, namely that you not only disagree with my work, but actively discountenance it in a way that infallibly reacts on me personally'.[12]

Rivers continued to work both as anthropologist and as psychologist. In anthropology, he remained faithful to the diffusionist position he and Elliot Smith had developed before the war. Indeed, new evidence that the mortuary customs of the Solomon Islands were more or less identical with those of ancient Egypt provided Rivers with the final proof that the culture of the early dynasties of Egypt had been carried far into the east. To both lay and specialist audiences he hammered home that message.[13]

Rivers' enhanced stature and new-found willingness to accept office meant that he suddenly became a powerful public figure. In 1921 he was president of the Folklore Society, the psychology section of the British Association for the Advancement of Science and the Royal Anthropological Institute. He and Elliot Smith now began to hatch plans to use this power base as a springboard for 'ethnological propaganda', pushing through their diffusionist agenda.[14]

At the end of 1921 there came a further development. Rivers, who had become increasingly involved with adult education and Labour Party intellectuals, was invited to stand as the Labour candidate for the University of London parliamentary seat. At first he hesitated, but in February 1922 he accepted – and at once delivered a long lecture on 'Psychology and politics'. What led him to take this step? Although he joked about psychoanalysing Lloyd George and in his lectures warned against crude or premature attempts to use psychology's 'supposed discoveries' in politics, standing for Parliament was a natural result of Rivers' own change of mood – and of the failure to establish the 'new psychology' in Cambridge. Besides, two of his medical mentors had represented the London University seat. This extra duty meant, however, that he was now overwhelmed with work. Frederic Bartlett observed Rivers at this point: 'I saw him all alone; at breakfast with his friends; at lectures; in discussions of the Psychological Society; at his squashes; sitting at the table at the College Council, and sometimes every other impression would vanish away before the sudden overwhelming impression that he was horribly weary.'[15]

On the morning of Saturday 3 June 1922, Rivers let his college servant off for the Whit weekend. Then, during the night, he developed an acute intestinal problem – a strangulated hernia, the twisting of the bowel. He lay for many hours in terrible pain until he was found the following day. He was taken to hospital and operated on but it was too late to save him. Rivers died on Sunday 4 June. He was only fifty-eight.[16]

'Never have I known so deep a gloom settle upon the College as fell upon it at that time', Bartlett wrote a year later. St John's struggled to organise a suitable funeral. Siegfried Sassoon, who had seen a good deal of Rivers since the war, came up by train from London with Henry Head, Grafton Elliot Smith and 'a little quiet anthropologist called Perry'. He listened as Elliot Smith and Head told stories about

their friend: 'How serene and wise they seemed! They made me feel
that the funeral would be quite easy to get over; there seemed no
cause for lamentations and regrets. Only profound gratitude to the
dead man for all that he wrote and lived.' It was a brilliantly sunny
day in Cambridge, the first day of the May Week races.* Sassoon's
party went up to the rooms Rivers had occupied for ten years and
Head recalled their collaboration in the 'nerve experiment'. Then they
made their way to St John's chapel for the service. Sassoon was 'without
emotion or apprehension of emotion' as he sat down.

> The single bell tolled at long intervals. Head whispered 'the organ is
> out of gear but Rootham has arranged for some beautiful choral music'.
> There was a hush. And then, while we all stood up, from the college
> courts there came a sound of distant singing. As the procession entered
> the chapel, the singing swelled to a joyous triumph. It was all up with
> me as soon as I heard that music. I did struggle hard to remain unmoved.
> But all the repression of the past two days collapsed. And that was the
> end of my self-control.

At the end of the service, the procession carried Rivers' coffin as far
as the college gate; then it went off to the mortuary, to be cremated.[17]
    That evening, Sassoon and Henry Head, now back in London,
discussed the meaning of death. Elliot Smith, however, had other
concerns. 'Rivers' death is a real catastrophe and compels me to make
new orientation in life, for it has upset so many plans that we were
developing', he wrote to Perry.[18]

---

* By then, May Week, the traditional period of festivity at the end of the Cambridge
summer term, took place in June.

'The priest who slew the slayer'

Those trees in whose dim shadow
The ghastly priest doth reign
The priest who slew the slayer
And shall himself be slain.
    Macaulay

In 1919, when Grafton Elliot Smith's reputation as a neuro-anatomist and teacher of anatomy was at its height, he was headhunted by University College London. Already receiving substantial sums of money from the Rockefeller Foundation, the college needed a major figure to spearhead expansion. Elliot Smith at once hatched his post-war project – a huge new department of human biology, in which anatomy, physiology, psychology, anthropology and archaeology would all be taught together. This dream was not fully realised because the psychologists and archaeologists at UCL refused to play ball, but Elliot Smith built up an empire which included a professor of medical history and a readership in anthropology, tailor-made for Rivers. When Rivers declined to leave Cambridge, the post went to William Perry.[1]

Elliot Smith's longstanding interest in the evolution of the human brain inevitably drew him into the early history of man. He took part in an argument over the mechanism by which our primitive primate ancestors were transformed into human beings. Was the decisive stage the shift from trees to the ground (terrestriality); the change of posture from walking on four legs to balancing on two (bipedalism); the expansion of the brain, with flowering of intelligence and language (encephalisation); or the emergence of technology, morals and society (civilisation)? Elliot Smith argued that the expansion of the brain came before all else, and when his former assistant Frederick Wood Jones took a different view, his brutal attempt to suppress Wood Jones's dissenting opinion led to the breakdown of an old friendship.

Elliot Smith was also drawn into palaeontology – the attempt to reconstruct man's prehistory from his fossil remains. At this time, it was not easy to date fossils and the human fossil record consisted primarily of Neanderthal Man, first found in 1856 in Germany, who seemed in many ways a remote relative rather than direct ancestor of man; Java Man, found in the 1890s; and a small number of French fossils. Asia was generally considered to be the 'cradle of mankind'. Britain, it seemed, did not feature in prehistory at all.

There was therefore enormous excitement when it was announced in 1912 that a Sussex solicitor called Charles Dawson, assisted by a young French Jesuit, Pierre Teilhard de Chardin, had found, in a gravel pit at Piltdown near Uckfield, fragments of a human brain-case, jaw and teeth, lying in geologically ancient sediments. After reconstructing the skull, Arthur Smith Woodward of the British Museum proposed that Piltdown Man represented an evolutionary missing link between ape and man. As the leading expert on the development of the brain, Grafton Elliot Smith was brought in, and pronounced Piltdown to be genuine: that the jaw and cranial fragments belonged to the same creature had never been in any doubt on the part of those who had seriously studied the matter, he declared, before he had himself examined the finding.[2]

In fact the Piltdown fossil was a forgery, a composite of three distinct species, consisting of a human skull of medieval age, the 500-year-old lower jaw of a Sarawak orang-utan and the fossil teeth of a chimpanzee. Elliot Smith was not the forger (as would later be implausibly alleged); in fact his involvement was hasty and marginal. Heavily preoccupied with teaching in Manchester, the development of his theory of cultural diffusion and (after 1915) wartime medical work, he became embroiled in a bitter argument with his friend Arthur Keith about the accuracy of the reconstruction of the Piltdown head carried out by Smith Woodward – a row which did much to divert attention from the more fundamental argument about the authenticity of Piltdown Man. Writing to Keith in September 1913, Elliot Smith admitted that the British Museum's reconstruction was not perfect. But 'why magnify these discrepancies in the lay press and before a number of foreigners'? Keith ignored this advice and publicly attacked

Elliot Smith's views at the Royal Society. 'It was a crowded meeting,' Keith recalled in his autobiography, 'and it so happened that he and I filed out side by side. I shall never forget the angry look he gave me. Such was the end of a long friendship.'[3]

Why was Elliot Smith, along with most of the leading British authorities, so easily taken in? Why did he tell the British Academy in 1916 that 'the discovery of the remains of the Piltdown man is perhaps the most remarkable episode in the whole history of anthropology'? Piltdown has become the classic illustration of how conclusions are sometimes reached in science in spite of, rather than because of, the evidence. British palaeontologists were desperate to get one over on their continental rivals. But more importantly, the combination of a human-like cranium with an ape-like jaw tended to support the notion then prevailing in England that human evolution had begun with the brain. 'The outstanding interest of the Piltdown skull', wrote Elliot Smith, 'is in the confirmation it affords of the view that in the evolution of man the brain led the way. Man at first was merely an Ape with an overgrown brain.'[4]

It was a tragedy that Elliot Smith was led astray by Piltdown Man. It prevented him from grasping fully the significance of other findings, most notably those of Raymond Dart, a young Australian who joined Elliot Smith's team at University College in 1922. Dart had high hopes of getting a senior post in Britain and was not best pleased when Elliot Smith (who made a habit of arranging his students' futures) informed him that the chair in anatomy in Johannesburg had been procured for him. With some reluctance, he went out to South Africa.

In fact, Dart was extraordinarily lucky: two years after his arrival he heard of a limestone cliff at Taungs in Bechuanaland which might contain fossil remains of primitive man and sent colleagues to bring back rock samples from it. In 1925, without bothering to alert Elliot Smith, Dart published in *Nature* a short paper announcing that he had found 'a specimen of unusual value in fossil anthropoid discovery' which showed evidence of 'an extinct race of apes intermediate between living anthropoids and man'. Dart's find, which he christened *Australopithecus africanus*, was the skull of a child with the face of an ape but the brain of a human – not in size but in some elements of its architecture. Elliot Smith saw the importance of Dart's discovery but the difficulty of reconciling it with Piltdown prevented him from endorsing it completely; and so, along with most British experts, he refused to accept that there was anything human about Dart's fossil. The orthodox view remained that man had originated in Asia, probably in the Himalayas. Today we know that the 'Taung child' opened a new era in palaeoanthropology in which it came to be accepted that Africa was the cradle of mankind, just as Darwin had predicted.[5]

In the 1920s the range of Elliot Smith's activities broadened yet further. He began to write popular books about the ancient world, gave broadcasts on the BBC, and pontificated about Tutankhamen – but, despite energetic lobbying, was not asked to carry out the autopsy on the pharaoh's corpse. He was knighted in 1933. Such was his public standing that his students came to occupy chairs of anatomy all over the world.

It was largely left to William Perry to continue the argument for cultural diffusion, though Elliot Smith also used his position as William Rivers' literary executor to publish five volumes of posthumous works by Rivers, in which his support for the doctrine of cultural diffusion was emphasised. 'Professor Smith', a reviewer noted, 'shows a keen desire to enlist all the prestige of Rivers in support of the Smith–Perry thesis of the origin of all civilisation in the Ancient east and its diffusion therefrom throughout the world.' The first volume, though entitled *Psychology and Politics*, contained a long and eulogistic account by Rivers of Elliot Smith's work, annotated by footnotes in which Elliot Smith commended Rivers' flattering account of him. This led to some fierce exchanges with Alfred Haddon, who was trying to cleanse Rivers' legacy of the diffusionist heresy.[6]

At the same time, Perry now began to put flesh on the diffusionist programme. In 1923 he published *Children of the Sun. A Study of the Egyptian Settlement of the Pacific*; other works followed. Three years later, the popular writer Harold Massingham brought these ideas closer to British readers, arguing in his book *Downland Man* that the barrows and henges of Wessex had been built by immigrants from Spain intent on recreating the sun-worship cult of Egypt. Beautifully written and handsomely produced by Jonathan Cape, *Downland Man* was a bestseller.

In seeking to understand how the belief in immortality had developed in Egypt and how the Egyptians had sought the 'givers of life' around the world, Perry was not afraid to speculate boldly; the fact that cowrie shells resembled the female vulva, for example, was pressed into service. His vision of how a warlike caste had imposed civilisation upon a peaceful world was coloured by the pacifism and pessimism of the 1920s. But his claim to have discovered a new anthropology enraged the academics. Professor Peet of Liverpool 'could not refrain from astonishment', found 'amazing logic' and 'statements which will astonish the Egyptologist', and noted that Perry was 'not squeamish about chronology'.[7]

Professional archaeologists and anthropologists had long ceased to take Elliot Smith seriously. The death of Rivers had removed his only respectable ally and, in addition to Haddon and Flinders Petrie, he and Perry now had to contend with a formidable new enemy – Bronisław Malinowski, Seligman's young Polish protégé, who had accompanied Rivers and Elliot Smith to the British Association meeting in Sydney in 1914. When war broke out, Malinowski, though technically an enemy alien, was allowed to stay in Australia and, as a result, had spent the period of hostilities doing fieldwork – first, abortively, in New Guinea; then, heroically, for two years in the Trobriand Islands, further to the east. In 1919, he returned to Britain to write *Argonauts of the Western Pacific*, probably the single most revolutionary work in the history of anthropology. By spending years with his subjects and learning their language Malinowski set new standards in fieldwork and discovered that Rivers' famous genealogical method produced 'dead material, which led no further into the understanding of real native mentality or behaviour'. Malinowski regarded diffusionism as 'phantastic' and was only interested in current ethnographic practices. He completely rejected the idea that it was possible to deconstruct primitive peoples' history or place in evolution from their surviving customs and patterns of social organisation. Malinowski's ideas soon prevailed. His quickly became the only game in town: lecturer, reader and, by 1927, professor at the London School of Economics.[8]

For all his genius, Malinowski was an egomaniacal monster. 'Rivers', he once said, 'is the Rider Haggard of anthropology. I shall be the Joseph Conrad.' He attacked Elliot Smith and Perry without mercy and soon posed a threat to the anatomist's ambitious plans to use Rockefeller money. Early in 1927, when Elliot Smith learned that the Rockefellers intended to transfer their support to Malinowski, he sent a series of despairing, overwrought letters to New York. Perry, he declared, was 'the Galileo of this movement which within the next decade will effect as profound a reform in humanitarian studies as Galileo effected in physics'. But the feisty colonial upstart had now begun to be regarded as a pathetic, deluded old man. The Americans continued to make substantial grants to Elliot Smith's department of anatomy but transferred their funding for anthropology to the LSE.[9]

It was Perry who felt the strain first. Always delicate, he had begun to develop multiple sclerosis. 'His condition is very alarming and I doubt whether he can carry on much longer', Elliot Smith wrote on 9 June 1932. 'He, however, will not admit that he is not fit to undertake the work of next session and we are thus placed in a dilemma of extraordinary difficulty and delicacy. However, there is no evading the issue that it will soon be physically impossible for him to come to the College.' Somehow, though, Perry, struggled on.[10]

Then it was the 'old man's' turn to weaken. One day in 1934, following a discussion on Peking Man at University College, Elliot Smith and his South African protégé Solly Zuckerman were returning to his home in Primrose Hill. As they waited for a bus at Camden Town, Elliot Smith took out a handkerchief and, wiping the side of his nose, remarked strangely; 'I am wondering which blood vessels in my brain are leaking. All day I have felt as though I've a cold on the right side of my nose and there's been a slight tingling in my fingers. I wonder how serious a stroke it's going to be.' Soon afterwards he did indeed suffer a stroke, from the effects of which he would never recover. Two years later he was pushed out of University College.[11]

By this time Elliot Smith had fallen out with both Arthur Keith and Frederick Wood Jones but the two continued to correspond about their old friend. Keith thought that Elliot Smith had 'a nose for what way the truth lies but [was] greedy for all the kudos available . . . he suffers from two bad and very incurable disorders: jealousy and egomania.' Wood Jones admitted to Keith that 'If Elliot Smith had refrained from his nasty methods with me I would not have been at the trouble to hunt out his own inconsistencies in his special field.' But, hearing of his misfortunes, they both felt guilty. Keith knew he ought to go and see the old man in the nursing home in Sidcup to which he was now confined but chose not to: 'we differ so radically we could only exchange nothings'. Replying from Melbourne, Wood Jones offered his own verdict:

> I am sincerely sorry for old ES; he seems to have come to a very sad old age . . . I fancy that at bottom all his troubles are the outcome of the fact that he is an Australian & that he married an Australian wife . . . He had a good brain, great charm, and a lot of Huxley's characteristics. But . . . he was at bottom, like so many Australians, lacking in those things, loyalty, sincerity, altruism – and the hundreds of other things, that constitute the basis of (hated word) a gentleman. I would have been very loyal to ES, for I learned to be very fond of him in Nubia – where he was at his best . . . in camp he was the finest companion one could wish for. But social ambitions, mainly due to his wife – intense jealousy of his fellows – intolerance – and a little devious streak, not altogether nice, made him hard to serve as a loyal master. Still I shall always owe him a debt for the days in camp in Nubia – and I am very sorry his old age is being a sad one.[12]

Grafton Elliot Smith died in 1937 in a nursing home in Broadstairs, aged sixty-five. Wood Jones wrote three obituaries of him and only in one did he hint at the breakdown of his relationship with his friend. 'He was a genial man, charming and courteous in debate, so long as the debate was carried on by the spoken word', he told the *Australian National Review*. 'But once the pen was in his hand the gentleness forsook him . . . The pen is doubtless mightier than the sword, but, all too often, Elliot Smith employed it as a dagger.'[13]

McDougall in America

In 1919 William McDougall accepted the offer of the chair of psychology at Harvard University. Hoping to make a fresh start, exhausted by his wartime labours, shattered by the death of his younger daughter from rheumatic fever, and weary of the scorn of Oxford philosophers, he hoped to find in America a country where psychology was more welcome and he could step into the shoes of his hero William James.

At first all seemed to be going well. 'We are very comfortably settled in a convenient little house among very kind neighbours and everybody is very genial and helpful. My wife, as you may suppose, is greatly appreciated and I think she will be quite happy', McDougall wrote to the physiologist Charles Sherrington. His boys liked their school; his daughter Lesley was being instructed in painting by William James's widow, and McDougall found a summer retreat at Silver Lake, New Hampshire, where he produced two solid books of academic psychology. 'Funds for work are not as copious as I had hoped', he wrote, 'but that is the only way in which things at all fail to come up to my optimistic forecasts. So I have good reason to be content.'[1]

In fact, things had already started to go wrong. On arriving at Harvard, McDougall had been invited to give the Lowell Lectures and, in his eagerness to make an impact, made a serious misjudgement. The lectures he delivered in Boston were a revised version of a talk on anthropology and history given in Oxford a few months earlier: a provocative but intellectually patchy discussion of *European* history in terms of racial characteristics, of the sort that T. H. Huxley had delighted in and which Oswald Spengler and Arnold Toynbee would spin into best-selling books. On to this, however, he tacked a sensationalist title, 'Is America Safe for Democracy?', and a preface prophesying the imminent destruction of the American nation should it continue on its present path to racial equality. Most contentiously, when discussing the innate mental development of different groups, McDougall widened his argument to include America's black population – 'though I would rather choose for discussion any other race than the Negro', he declared, 'they alone of the coloured peoples have been studied in a way which makes possible a comparison with the white population'. He then asserted that blacks were of lower intelligence, using as evidence data from the intelligence testing which psychologists had carried out for the US Army during the war; and that they were submissive by temperament, a claim he based on purely anecdotal and subjective material.[2]

McDougall was stepping into a minefield. For over a decade, a debate had been raging in the United States over questions of race and immigration, and psychologists had become intimately involved in it. Intelligence-testing techniques developed by the French for helping backward children in education had been modified by American psychologists to produce instruments such as the Intelligence Quotient (IQ) which could measure intelligence in the general adult population, primarily in order to see whether immigrants arriving in America were intelligent enough to prosper there – the implicit purpose being to exclude less intelligent arrivals from southern and eastern Europe. Then, during the war, America's psychologists (led by McDougall's friend, Robert Yerkes) had persuaded the US Army to let them test the intelligence of military inductees, in order to obtain data on the relationship between race and intelligence. Using methods which were deeply flawed – mainly because they reflected the cultural assumptions of America's Ivy League elite – they had concluded that black Americans were less intelligent.[3]

By drawing on this study, not only did McDougall commit serious methodological mistakes – for example, he ignored Yerkes's warnings that the army group test was 'a dangerous method in the hands of the inexpert. It was not prepared for civilian use, and is applicable only within certain limits to other uses than that for which it was prepared' – he also climbed into bed with undesirable bedfellows. Although he had tried to strike a middle position between the outright racists in the Madison Grant tradition and those who did not believe there were any distinctions of race at all, such subtleties meant nothing in the polarised climate in the United States. McDougall now found himself lumped together with figures such as Theodore Lothrop Stoddard, author of *The Rising Tide of Color against White World Supremacy* and, in time, supporter of Hitler. McDougall later recognised that the lectures had been a huge mistake. 'I did not then realise that in touching, however impartially, the racial question, I was stirring up a hornet's nest', he wrote. Although such views were quite common among America's psychologists at this time, most of them did not broadcast their opinions as he had done. His views on race were partly responsible for his failure to attract a following among graduate students at Harvard.[4]

Even if McDougall had kept his mouth shut on the subject of race, he would have faced other difficulties in America. His arrival in Boston coincided with a major sea change in American psychology which meant that most of his Harvard colleagues were hostile to him from the start. Psychology at Harvard was still part of the philosophy department and both disciplines had fallen on hard times in the decade following the death of William James in 1910. In philosophy, the university's president, Lawrence Lowell, excluded outstanding candidates such as Bertrand Russell, Arthur Lovejoy and John Dewey as being socially or morally unacceptable and appointed lesser people instead. Among the psychologists, terrible internal wounds festered. Back in the 1890s, when William James had decided to move from psychology into pure philosophy, he had gone to Germany and personally recruited Hugo Münsterberg, the rising star of German experimental psychology, to be his successor in the laboratory. Münsterberg was a brilliant and energetic researcher who quickly branched out, seeking to apply psychology to American life and culture and to the needs of industry and commerce; he also wrote a pioneering work on the cinema, America's infant art form. Münsterberg's vulgar manner and entrepreneurial ventures alienated his Harvard colleagues but they put up with him – until the Great War began. Despite his Jewish background, Münsterberg was a fervent German patriot, whereas most of Harvard was passionately pro-British; but instead of returning to Germany, Münsterberg opted to stay in Cambridge and continue teaching. His life was made increasingly difficult and eventually the strain proved too much. During a lecture in December 1916, he collapsed, fell from the podium, and died.[5]

Harvard was desperate to find a big name to replace him. Yet none of the obvious American candidates proved suitable. 'In this situation,' writes the historian Bruce Kuklick, 'the philosophers had William McDougall appointed to the psychology professorship in 1920. The philosophers wanted a man with a name and a theoretical and philosophical orientation. McDougall and his books were world famous.'[6]

McDougall was not the man for such a situation: his lifelong ambiguity towards groups and institutions had not equipped him to provide leadership in a crisis. His real problem, though, was that by the time he got to America his brand of psychology had become profoundly unfashionable. The climate in American psychology was now experimentalist and positivistic, whereas McDougall was in the older, more philosophically oriented tradition. His instinct theory, for example, had been new and exciting back in 1908, but now instinct theories were two a penny in psychology, though no one could agree on how many instincts there were. Where McDougall had produced an inventory of eleven, other texts listed seventeen, twenty-six or even 110. The pendulum had swung, and now the way to make a name for oneself was not to come up with a theory of instincts but to attack the whole doctrine itself. 'Are there any instincts?' asked the first such assault, launched in 1919; many others followed. Nor did it help McDougall's standing that when *The Group Mind*, his long-awaited follow-up to *An Introduction to Social Psychology* and the book he had brooded over throughout the war, appeared in 1920 it was generally regarded by American academics as a work of opinion rather than of scientific substance. 'The author is really voicing in large part a set of value judgements which have become habitual and the justification for which he has largely lost sight of, if indeed he ever subjected them to a thorough-going criticism', the political philosopher George H. Sabine wrote. Other American reviewers called it 'a disappointment' and 'a contribution to idealistic philosophy rather than to collective psychology'.[7]

McDougall's fiercest critic was the *enfant terrible* of American psychology, John Broadus Watson, the advocate of a new movement known as 'behaviorism'. Born in 1878 and raised on a small farm in South Carolina, Watson had somehow overcome his background to study philosophy at the University of Chicago and write a PhD thesis based on experiments with rats. Appointed professor of psychology at Johns Hopkins, Watson made his name with his address 'Psychology as the Behaviorist Views it', given at Columbia in 1913. So far, said Watson, psychology had completely failed to establish itself as a natural science because it had long been concerned with the study of the mind, especially with consciousness. Such an enterprise was doomed to failure. It must always be subjective and unscientific, especially in the whole quasi-philosophical business of 'introspection'. The only way to make psychology scientific was to confine itself to externally observable objective phenomena, to study human behaviour as that of animals was studied. 'Psychology as the behaviourist views it is a purely objective branch of natural science. Its theoretical goal is the control and prediction of behaviour', Watson declared.[8]

Behaviourism was not a complete system. It did not provide a convincing account of language, let alone of thought. In explaining human behaviour Watson drew very heavily on the researches of the Russian physiologist Ivan Pavlov; humans, like animals, learn by conditioning, he argued. Thoughts, feelings and intentions, indeed all mental processes, do not determine what human beings do. We are biological machines and do not consciously act; rather we *react* to stimuli. Consciousness, mind and mental states were therefore to be ignored. This was an updated version of what McDougall had consistently denigrated – mechanistic psychology. To Watson, on the other hand, McDougall's theory of instincts was a relic of Victorian times: something dreamt up in an armchair.

For many young psychologists in America, Watson was a hero, even though a divorce scandal had forced him to leave academe for the advertising industry in 1920. His attacks on McDougall eventually forced the latter to respond and in February 1924 a full-scale debate took place in the Psychological Club in Washington. Here Watson repeated his view that anyone who introduced consciousness into psychology did so because of spiritualistic or vitalistic leanings. 'The behaviorist cannot find consciousness in the test tube of his science', he declared. 'He finds no evidence anywhere for a stream of consciousness.'

McDougall responded that 'the behaviorist knows nothing of pleasure and pain, of admiration and gratitude. He has relegated all such metaphysical entities to the dust heap, and must seek some other explanation.' However,

Dr Watson and I are engaged in the same enterprise, the endeavour to reform psychology by correcting the traditional tendency to concentrate upon the facts of consciousness to the neglect of the facts of behavior. The difference between us is that I, unlike Dr Watson, have not made myself at the same time famous and ridiculous by allowing the impetus of my reforming zeal to carry me over from one lop-sided position to its opposite, from exclusive concern with the facts of consciousness to exclusive concern with the facts of behavior.

McDougall narrowly won the vote that followed, although he claimed that his margin of victory would have been much greater had the women in the audience not all voted for the handsome Dr Watson. But, in the eyes of his professional colleagues, Watson had been the clear winner. The Englishman had been 'knocked out of the ring' in the 'Battle of Behaviorism'.[9]

If McDougall's system of psychology,* with its emphasis on instinct and purpose, left him isolated from his Harvard colleagues, his interest in psychical research embarrassed and scandalised them. At the age of fifty, he had long since ceased to feel the need to justify his interest in this field, which anyway had recently been reanimated by the Great War, with thousands of parents on both sides of the Atlantic now trying to make contact with sons lost on the battlefield. Shortly before his departure for America, McDougall had been chosen as president of the English Society for Psychical Research, so it seemed quite natural for him to assume the presidency of its American sister body in May 1921. He also took part in public events: when in 1922 *Scientific American* magazine offered a prize of $2,500 for the first satisfactory photograph of a psychic phenomenon, McDougall agreed to sit on the committee of experts which assessed claimants (the escapologist Harry Houdini was another). There were, however, significant differences between the two organisations. Whereas in Britain psychic

---

* McDougall's school was generally known as 'Purposive psychology' but by 1930 he had rebranded it as 'hormic psychology' – from *hormé*, the Greek spirit of activity and initiative, a term which had already been used in a different way by Jung. 'Hormic psychology' proved to be an unfortunate coinage and did not catch on. W. McDougall, 'The Hormic psychology', in C. Murchison (ed.), *Psychologies of 1930* (Worcester, Mass., 1930).

research was socially and intellectually respectable – controlled (in organisational terms) by Mrs Eleanor Sedgwick, sister of the prime minister A. J. Balfour and widow of a prominent Cambridge philosopher, and with figures such as the writer Arthur Conan Doyle and the scientist Sir Oliver Lodge publicly associated with it – in the United States the movement was less genteel, more low church and less interested in scientific enquiry. American psychologists were quite happy to take money given by benefactors for the purposes of research in this field, but they were careful to operate in the psychological laboratory, rather than the seance parlour.[10]

When McDougall voiced his suspicions that 'Margery', a spirit medium and a claimant for the *Scientific American* prize, was a fraud, a rift opened up. Powerful figures in the American Society for Psychical Research not only believed 'Margery' to be genuine, they suspected McDougall of being unsound on the survival hypothesis – the assumption that mind and personality survive the death of body and brain – and too intellectual in his approach. 'Spiritualism appeals to all classes of the community', one of his opponents remarked. 'Hence the majority of its adherents must needs be persons who lack the capacity to understand and to appreciate the necessarily difficult methods of analysis and argument on psychical research.' In a well-organised coup, the membership ousted McDougall from the society's presidency and replaced him with a more populist candidate. McDougall and his supporters responded by founding a rival association, the Boston Society for Psychic Research.[11]

All of this made good copy: 'Professor McDougall' was soon a familiar figure to *New York Times* readers and, according to a modern study of that newspaper's coverage, 'came to be seen as a comic figure, one who was not to be taken seriously'. This type-casting, once established, was not easy to shake off: 'From the *Times*'s point of view, he was involved in sensational, shady, not-quite respectable topics, such as drugs, eugenics and extrasensory perception.'[12]

Nor would McDougall's professional standing have been enhanced by the programme of research he embarked on at Harvard. With his usual contempt for established scientific orthodoxy, he had long questioned the standard Darwinian account of how inheritance works and had flirted with the rival version, first propounded by Jean-Baptiste Lamarck in the 1800s, according to which acquired characteristics can be inherited. McDougall had several times proposed that some major scientific body should carry out 'a prolonged experiment designed to settle the Lamarckian question once for all, using preferably dogs as the most likely material', but no one had taken up the challenge. Now, at Harvard, he set to work himself but, realising that an experiment with dogs would take too long, used fast-breeding white rats instead. 'The question at issue seemed to me the most important question yet formulated by the mind of man and clearly susceptible of solution by experimental procedure', he wrote.[13]

McDougall placed his rats in a water maze. 'It was a two-choice apparatus,' a student later recalled, 'the animal being put in a tank of cold water, having then to swim to the right or the left, to choose between a lighted ramp or a dark ramp by which to escape from the water on to a dry platform. If he chose the wrong ramp he received a walloping shock . . . some animals did the task with a fair amount of swimming back and forth to "con" the two ramps before choosing, whereas others simply turned one way and charged the ramp, whether it happened to be lighted or not.' The question then was whether the rats which got it right would pass that knowledge on. 'McDougall counted the number of times a group of rats was placed in the tank before they learned to escape by the dim gangway; he then inbred that generation and tested its descendants to see whether the new group of rats would learn more quickly to avoid the bright gangway.' By 1930, when ten years had passed and twenty-five generations of rats had been in the maze, the answer was still unclear.[14]

Then, finally – on top of the racial theories, the involvement in psychic research and the Lamarckian experiment – people were put off by McDougall's manner: his lordly English way of talking, his habit of patronising Americans, the sense he gave of 'still thinking of himself as an Englishman in the British Colonies'; his Oxford tweeds.[15]

Most American psychologists didn't bother to read McDougall's books; they took their picture of him from the media. Quite quickly, he not only became completely isolated from his Harvard colleagues but a demon figure across the profession. 'An intense wave of anti-McDougall . . . feeling swept through American psychology', a friend later remembered; his professional colleagues, another wrote, 'looked upon him not only as an outsider but as a foreigner who wished to impose his views upon them'. A pack instinct was aroused. McDougall began to be belittled and treated disrespectfully on public occasions. At the Ninth International Congress of Psychology, held at Yale in 1929, the chairman launched a savage personal attack on him.[16]

His pride wounded, McDougall began to look for other positions. In 1925 he talked of going to Berkeley and applied for the chair of mental philosophy and logic at Cambridge, England, but his old university didn't want a psychologist and appointed G. E. Moore (and later Ludwig Wittgenstein) to the chair. For the moment McDougall stayed at Harvard. 'My first enthusiasm has a little waned', he told his friend, the philosopher Samuel Alexander, in May 1926. 'I begin to think of returning home to settle down.'[17]

But then McDougall was approached by William Preston Few, the president of Duke University in Durham, North Carolina, who had embarked on an ambitious scheme for transforming what had until recently been a small Methodist training college into an Ivy League university for the South, funded by the millions of the Duke tobacco and construction dynasty. President Few was looking to appoint a professor of psychology. McDougall advised him against several candidates whom he regarded as unsuitably behaviourist and then – finding that Few shared something of his own outlook – surprised him by putting forward his own name. Negotiations followed. McDougall secured a salary of $9,000 and the promise of a journal, a laboratory (in which to pursue his Lamarckian experiments) and staff; in return President Few obtained the prestigious head of 'one of the top ten psychologists in the world'. Money apart, Dr Few was careful to feed McDougall's vanity, having been told by the dean of divinity at Yale that McDougall was 'a reasonably good yoke-fellow if you give him his daily meed of praise'.[18]

McDougall hesitated briefly. His old friend Charles Sherrington, now president of the Royal Society, asked him to return to England and work with the Medical Research Council to 'try to do something about the deplorable state of psychiatry in Britain'. But as no firm university chair was involved, McDougall turned it down. 'We have decided to "bury ourselves" as some will say in North Carolina', he wrote to Sherrington on 22 December 1926.[19]

When McDougall arrived at Duke in the New Year, he found the place a building site, the new campus, in the 'Tudor-Gothic' style pioneered at Princeton, still rising. But Dr Few kept his promises. McDougall was quickly able to appoint the staff he wanted, get his rat experiments going again, establish a journal (*Character and Personality*) and even to create a programme of psychic research. But the disadvantages of the place were more cultural than physical. As McDougall quickly discovered, 'a good many of the students [were] rather crude persons' and Durham was 'an unattractive city . . . completely lacking in amenities'. He would not have minded the fact that black students were not admitted (and would not be until 1963), but the strongly religious climate came as an unwelcome surprise. '*Eruditio et religio*' was the university's motto. At Harvard freedom of enquiry was an established fact; here it was a conditional one.[20]

The culture McDougall now inhabited was revealed in the torrent of letters which he received asking for advice. A New York teacher preparing a talk for the Students' Science Congress on 'Remaking Man' wanted to know what psychology could do to remake man – if possible, by return of post. A superintendent of schools asked 'when is a girl socially efficient' (McDougall confessed he had no idea). A lady from Springfield, Ohio, wanted to know what books on psychology he would suggest 'that a young couple have in their library to enlighten them on the following subjects: children; understanding of our fellow men; getting along with our fellow men'. Mostly, McDougall showed great patience, referring the medical cases to a reputable psychiatrist living nearby, but he lost his cool with a woman who wrote to tell him that, contrary to his assertion that there were twelve instincts, there were in fact thirteen.* 'I find it very difficult to convince anybody in this country of any instinct whatsoever and the possibility of convincing them of a thirteenth instinct is very remote', he replied.[21]

America was losing its charm. McDougall began to go to England for the summers. 'My wife has had enough of America and I must act accordingly', he wrote to Samuel Alexander, in July 1930. But his efforts to get a chair in England were unavailing. He turned down University College London because its laboratory was a 'desperate hole' unsuitable for his beloved rats; Oxford, because it didn't offer him enough money. In 1934, McDougall resigned from

---

* In the twentieth edition of *An Introduction to Social Psychology*, published in 1926, McDougall added a twelfth instinct – that of laughter.

all the learned societies in America and made a serious attempt to
return to England, but he was persuaded to stay when President
Few allowed him to become a part-time professor on a full-time
salary. His need of Duke's money was all the greater because he
had lost heavily in the stock market crash of 1929 and been swindled
over an oil well by two confidence tricksters. The American press
enjoyed his misfortune:

It was the Rats that Set the Trap
New York police recently nabbed two men charged with swindling
Professor William McDougall, world-famed psychologist, out of $22,800
in a fake oil well. The professor is noted for his experiments on animal
behavior through tests made mostly with rats. Seems this time he got
caught by two oily ones. Slippery beasts to handle. Note: even absent-
minded professors can get a good grip on reliable business opportunities through the want Ads. That is how this one found a renter.

His professional enemies greeted this news with gleeful *Schadenfreude*;
as a friend noted, 'to the general reader of newspapers, a psychologist who allows himself to be deceived thereby discredits his
science'.[22]

There was also another personal tragedy. McDougall's eldest boy,
Duncan, was a cheerful extrovert. At Dartmouth College he sat an
intelligence test: 'I got thirty out of a possible hundred,' he wrote to
his mother, 'I always knew I was stupid.' Duncan had trouble finding
a career; eventually, after picking fruit and working as a lumberjack
in the west, he reluctantly acceded to his father's wishes and joined
the civil service of Rajah Brooke in Sarawak. 'One has to bubble over
a bit once in a while and if not to one's mother, to whom can one
bubble?' he wrote to his mother, on the boat going over. 'I must try
to be the "strong and silent" man and suffer strongly and silently.' His
father saw it differently. 'He seems to like the work and in about ten
years he should be ready to produce a new edition of *The Pagan Tribes
of Borneo*', McDougall wrote to Charles Sherrington. But by 1928
Duncan was back in Britain, as a fighter pilot in the Royal Air Force.
On 6 May 1932 his aircraft hit the runway during a display at Hendon
and he was killed instantly. 'No one who talked to [McDougall] the
week his son . . . was killed in an air crash', a friend wrote, 'would

have guessed what a tragedy had befallen the man, who was greatly attached to his family.'[23]

In his popular writings, McDougall addressed many different topics: the 'race problem' in America; ethics, morals and citizenship in a post-religious age; internationalism and world peace; psychic research; and the place of psychology in the social sciences. But, to his great chagrin, he made little impact. His books tended to fall between stools. Most Americans still belonged to Christian churches and relied on the Bible for their moral guidance; the intellectual avant-garde embraced Freud; the psychologists were ardent behaviourists. McDougall was too English, too empirical, too conditional and too old-fashioned.

Yet his reputation revived when his chapter in the collective volume *A History of Psychology in Autobiography* was published in 1930. McDougall had been one of the first people to be approached when Carl Murchison of Clark University and Edwin G. Boring of Harvard conceived the idea of getting prominent psychologists to 'put into print as much of their personal histories as bears on their professional careers'. McDougall's contribution was one of the best things he ever wrote – sharp, lucid and revealing. Where other authors trotted out shopping lists of their publications and academic positions, McDougall produced a fine intellectual autobiography, fit to rank with those of Rousseau or John Stuart Mill, despite its brevity. He frankly acknowledged that his own arrogance had led him to take many wrong turnings in his life and wryly confessed his failure to sell a 'sane, all-round, well-balanced system of psychology' to the American public. 'The more I write, the more antagonism I seem to provoke', he noted. Comparing his own isolation with William James's 'magic touch which made all his readers his friends', McDougall concluded, 'I suppose it is that my uncompromising arrogance shows through, in spite of the taming it has undergone.' The 'autobiography' belatedly revealed to American psychologists that they had a big man in their midst. It 'earned him many admirers from among those who had been his critics', Edwin G. Boring, his enemy at Harvard, would later write. The chapter was 'so honest, so forthright, so simply true to principles that many at last discovered the *man*'.[24]

There was one area, however, where McDougall did not come completely clean. All his life he had yearned for recognition, for acknowledgement that he was a truly great man. Now, increasingly, he felt he deserved more than he had received:

> I have realised too late that I might have done more for my chosen science had I from the first spoken with a less modest voice . . . had I at the outset put forward my views in a more self-assertive and clamant fashion, I might have been acknowledged as the leader of a powerful and perhaps dominant school of psychology; instead of remaining a well-nigh solitary outsider playing a lone hand; I might even have 'put over' the type of psychology which I believe to be most nearly true, and to be indispensable for the advance of all the social sciences.

There was one quarter, in particular, from which McDougall felt he should have had greater appreciation – the psychoanalysts.[25]

In the past McDougall had been close to Jung. In 1914 he had expressed his admiration for the Swiss psychiatrist and announced that he was to have his dreams analysed by him. The war had then intervened and the trip to Zurich had to wait till 1919. After all the build-up it was something of an anticlimax. According to Jung, the analysis had been very brief because, though McDougall had provided a few dreams, he hadn't given Jung enough material to work with in interpreting them. McDougall's version was that analysis was not really possible 'for so hopelessly normal a personality as mine'. The encounter did not produce any great intimacy, but McDougall gave a sympathetic account of Jung's work in his Outline of Abnormal Psychology in 1926. There was no reciprocation, however. Writing to a friend before visiting America in 1932, Jung mocked McDougall's description of himself as 'hopelessly normal': the Englishman hadn't the faintest idea of 'the actual procedure of analysis' and was 'absolutely innocent of psychology'. This was an acute remark, for McDougall had completely failed to understand what a treacherous character Jung was, using others for his own purposes. In a later book McDougall referred sourly to the way that Jung was spending his time 'curing various millionaire American neurotics' instead of 'investigating the dreams of various primitive peoples', as he had promised to do. 'Jung has withdrawn himself more and more completely from contact and discussion with common mortals like myself. And the pronouncements which reach this world from the snow-capped Olympus on which he dwells may have been well calculated to sustain his old converts in the faith, but hardly of a nature to bring any new ones into the fold.'[26]

But it was towards Freud that the full range of McDougall's emotions expressed itself, his urge to criticise the Viennese sage continually in conflict with his craving for his recognition. As we have seen, in 1914 McDougall had been dismissive of Freud. But when Freud finally took notice of his views, by including a short account of *The Group Mind* in his *Group Psychology*, McDougall was delighted. Over the next decade, Freud began to develop a broader system of psychology, based less on clinical work with abnormal personalities and seeking to deal more with ordinary human experience. McDougall approved but felt that Freud should acknowledge more the similarities between his new thinking and McDougall's psychology. A confused exchange of letters with the English Freudian, Ernest Jones, in 1925 left Freud himself baffled; why, if McDougall wanted to be friends with the psychoanalysts, did he criticise them? 'Why did he write such things, if he is conciliatory and free of affect? There is something wrong here', he wrote to Jones.[27]

In July 1935, McDougall gave three lectures on psychoanalysis at University College London. 'Obviously proceeding from hurt homosexual vanity,' Ernest Jones wrote to Freud, 'their theme was how little you had taken his important work into account. You had appropriated his theory of suggestion without acknowledgement and had now given up all your back views, e.g. the Unconscious, the Oedipus Complex etc., reverting to a level of knowledge which he could have conveyed to you a quarter of a century ago.' McDougall still hoped to elicit some response from Vienna when he published the lectures, and put pressure on his publishers 'to get the book out before Prof. Freud should die, for I am quite anxious for him to see it. He cannot last much longer though he is still active intellectually.' In fact Freud would outlive McDougall. The following February Ernest Jones sent Freud 'a very sad lucubration by McDougall which shows how thoroughly upset he is by your refusal to love him'.[28]

Jerome Bruner was one of McDougall's last students at Duke University. The professor who introduced him to psychology in 1935 struck him as 'worldly, rather plodding in lecture style, with bursts of wit, an impeccable and alien figure one could see walking the college paths in heavy tweeds even as winter turned to spring'. Having failed to escape from Duke, McDougall was now obliged to live out the last act of his life there.[29]

Throughout his time in America McDougall suffered from deafness in one ear but was able to go on lecturing and giving seminars. Early in 1938, however, cancer of the bowel was diagnosed and he underwent a colostomy operation in the hospital at Duke. It was April before he was allowed home, still weak; but he began to recover somewhat, in the care of his wife, 'an angel of compassion', and his son Kenneth. He corrected the proofs of his latest book, was able to go for short walks and instructed Kenneth in his Lamarckian work in the lab, which his son was going to take over: 'I am able to hope that he may clinch it, even if I do not live to see that stage achieved.' This was especially important because in 1936 a scientist in Edinburgh had announced that he had repeated McDougall's experiments with rats but had not replicated his findings. Confident of recovery, McDougall arranged to give his usual introductory lectures in October and made a serious effort to master lip-reading, by having his favourite poems read aloud to him. When his wife and son failed to encourage him, he suspected the worst. Soon afterwards, he was told that the cancer had spread to the liver and was therefore terminal.

His condition now worsened and he was in great pain. By a huge effort of will, and heavily laced with morphine and aspirin, McDougall got through the first lecture but was obliged to cancel the others. On 6 October, while administering morphine injections to himself every four hours, McDougall began composing his last work:

> The pain at times is certainly too severe to be borne by sheer fortitude, at least by a sensitive white man like myself; what a negro or a red-skin would say of pain it is interesting to conjecture, and I think it is a fair question, the pain being regarded as objectively as possible.
>
> At its most intense the pain seems to fill and to constitute the whole of consciousness in the shape of a dim white ball, every other conscious function being arrested.[30]

The sickness had opened McDougall's eyes to medicine's ignorance of pain. He realised that Henry Head's writings on pain, with which he had long been familiar, were totally inadequate. Why was more not known about how pain worked, and which part of the brain was engaged?

His condition also made McDougall assess his own life. What had he achieved? How far had he succeeded in his ambition 'to extend the bounds of human knowledge'? He reviewed all his contributions to science, from his work on muscles in 1897 to his latest modifications of the theory of repression. He had hoped to have fifteen years of uninterrupted writing after his retirement and to produce three more great works: a 'History of Psychology and Allied Thought', in which 'all the material would be pointed up to show convergence to the one tenable modern theory (my own)'; a new comprehensive but concise 'Social Psychology' and – the crowning achievement – a 'Social Psychology of the American People'. Now, none of these would be written.

Yet, in one respect he had been very fortunate:

> In relation to my wife and children I have enjoyed the most outrageous piece of good luck. I say *piece* because it all seems woven together like one tapestry . . . My wife has proved to be as lovely in every respect as she looks, and though no one of the children equals her, yet in their way each one is a perfect natural person, lovely in both physical and moral sense; and there has been among us no single serious quarrel or difficulty.

McDougall's agony was eased slightly by the letters he began to receive. His Harvard student Gordon Allport reminded him that 'great prophetic leaders' had often been abused by their contemporaries. 'There has been no leader of psychological thought to withstand the obscurity of materialism excepting you. You have been the pillar through the storm of mechanism and inhumanity.' Another student assured him that no one since Wundt had done so much for psychology. The Harvard physiologist Walter Bradford Cannon urged him to 'regain some of the vigor that was so characteristic of you'. And, indecently late in the day, St John's College in the other Cambridge offered him an honorary fellowship. 'There are lots of us who are deeply grateful for your work,' an embarrassed Frederic Bartlett wrote, 'although, with a mixture of English vice and English virtue, we don't very often say.'[31]

William McDougall died on 28 November 1938. Afterwards the tributes flowed; even old enemies such as Edwin G. Boring were gracious to his memory. But when Charles Myers eventually wrote to Annie McDougall, whom he had once known well, he found it hard to gush:

> Our thoughts and activities and views ran on rather different lines, but no one admired his outstanding ability and his marvellous energy more than I did. He achieved a greater success in spreading his ideas than he was ever willing to realise. I have just completed an obituary notice for the *Lancet* and realise more fully than ever what a wonderful life he led. Your loss is shared by many.

McDougall's last book, *The Riddle of Life*, had made a brief but cutting reference to Myers. But in the *Lancet* Myers was more gracious to the man he had once idolised and whom he had urged Alfred Haddon to take to the Torres Straits:

> His massive leonine face, his deep voice, his athletic figure and pride of body, his self-confidence and independence, and, above all, his exceptional mental powers marked him out as a man of outstanding personality. But he was a disappointed man and made few friends. He has indeed described himself as 'domineering' and 'uncompromisingly arrogant'. He has also said that he always found it difficult to believe in the value of his work, and that the more he wrote, the more antagonism he seemed to provoke. He may not have obtained all the reward and praise which his unusual ambition led him to expect. But the value of his work has not been surpassed by any psychologist of recent years.[32]

Myers and Industrial Psychology

When the Prince of Wales visited the National Institute for Industrial Psychology in the early 1930s, his car stopped short of the institute's building, Aldwych House. In their anxiety to get things right, the institute's director, Dr Charles Myers, and its president, Lord Macmillan, rushed along the street to take up new positions. In the confusion, however, the royal visitor was ushered through the wrong door and into the wrong lift, and his party ended up on the top floor of the next-door office-block. Unfortunately, irreverent juniors noted, neither Myers nor Macmillan had enough familiarity with the two buildings to realise that they were in the wrong one.[1]

The National Institute for Industrial Psychology certainly needed the royal seal of approval. Sixteen years after its establishment, it was still living from hand to mouth. The idea of bringing the insights of modern psychology to the shop floor did not originate with Myers – McDougall's predecessor at Harvard, Hugo Münsterberg, is generally regarded as the father of industrial psychology – but Myers was the first to give it concrete form in Britain. By 1917, when his interest was fired by a book on the subject, scientists were beginning to look at the effects of 'human factors' on industrial output, especially in munitions factories; finding, for example, that longer working hours did not necessarily increase output. A government body, the Industrial Fatigue Research Board, was also trying to devise ways to help workers doing repetitive and boring tasks. It concluded that 'constantly repeated movements were easier to perform if they were based on a constant rhythm'.[2]

In April 1918 Myers gave a speech setting out a possible role for psychology in peacetime industry. In the audience was Henry John Welch, a businessman who had similar interests. The two men quickly hatched a plan for an institute for the promotion of 'vocational selection and guidance', using scientific knowledge to improve conditions of work in factories and offices. Myers' vision was of using psychology to create a new kind of workplace, in which the harsh industrial climate of the pre-war years would be replaced by a new harmony. Rational-minded employers could, by the careful use of expert psychological advice, make their workers more contented and, in the process, enhance both productivity and profit. As Welch put it, 'We all look forward to a time when working people of all types in this country will be engaged in the work for which their temperaments and abilities most fit them, and when they will be able to return to their homes after their day's work is done, not too fatigued or disgruntled to interest themselves according to their inclinations and capacities, in literature, art, music and the higher things of life.' Myers took a year off from Cambridge to write *Mind and Work*, which set out his ideas on the subject. By 1921, the National Institute for Industrial Psychology had become incorporated and Myers realised he would have to choose between the institute in London and the psychological laboratory in Cambridge. After some agonising, he opted for London.[3]

The institute began life in 329 High Holborn, reckoned by those who worked in it as 'the ugliest building in London', with brown and white patterned tiles outside and in. Despite its imposing name, its staff consisted of Myers and one other psychologist. By 1930, however, the institute was housed in grander premises in Aldwych and employed fifty people, many of them psychology graduates. Their business, a member of staff later recalled, was 'rather sharply divided into "industrial work" and vocational guidance'. Normally the institute was called in by enterprises and organisations beset by such problems as absenteeism, high staff turnover or poor-quality work. The 'investigator', usually a psychology graduate with no experience of industry, would go off to York or Birmingham, talk to the workforce, and then offer recommendations which, surprisingly often, were accepted. In industries such as coal mining, engineering and confectionery it was possible to use 'movement study' to make monotonous physical labour more bearable, but throughout industry working conditions could be improved in quite basic ways, such as by redesigning the lighting. The reasons for high turnover of staff – for example, in Lyons' Corner Houses – could be explored through an 'opinion and attitude survey'.

One NIIP veteran thought the first group of investigators 'had "extraordinary success" – one can say "extraordinary" considering how few of them had had previous knowledge of industry or precedents to follow; but the combination of intelligence and theory worked in practice . . . the more enlightened firms were ones prepared to risk having this unknown species of adviser, the industrial psychologist'. She believed that the main strength of the institute lay in its down-to-earth approach. 'Its staff knew the smell of the back stairs of a factory and the problems associated with high labour turnover. They knew the kind of worries that beset people when they were in the wrong job or could not decide what career to follow. They also knew the excitement and pleasure of being able to solve these problems and having that help acknowledged.' The vocational side of the institute, initially run by Cyril Burt, spent its time devising aptitude tests for a wide variety of industrial jobs, from dressmaking apprentices to biscuit makers, and offering vocational guidance to individuals.[4]

After its first brave pioneering days, however, the institute hit choppier waters. The economic climate was not ideal, with British industry reeling from the hammer blow the war had dealt to textiles, coal and shipbuilding, the loss of foreign markets, and the arrival of new competitors. The expertise which Myers offered to employers was also available from more avowedly pro-capitalist competitors – whether it be Watsonian behaviourism or 'scientific management' as developed by Frederick Winslow Taylor. And by proclaiming itself neutral between capital and labour, the institute earned the distrust of both, British industry being wedded to the class system and not to be easily weaned off it. As a result, most of the institute's work came from a small group of enlightened and concerned employers, such as Cadbury's, Rowntree's and Debenhams. It did, however, provide the main source of employment for psychology graduates between the wars.[5]

In its second decade, the institute needed to go up a rank. But it never did. There began to be discontent in the ranks. Some felt that the institute needed to become better established and, in particular, to pay its staff properly; that bolder leadership was needed. Alec Rodger, a rising star, felt that the 'bittiness' of Myers' broad sweep caused problems: 'He had deep-rooted personal characteristics which

militated against his complete success in the role he had assumed.'
One of these was that Myers was an eclectic and 'disliked comprehensive theorising and "schools of thought" . . . Whether it was Freud
on the one hand or Spearman on the other his view was that people
with "theories" went too far.' This caution, Rodger argued, resulted
in a failure of leadership:

> His reluctance to draw firm conclusions and make pronouncements
> was great, and it issued in a hesitancy which could be dispiriting. A
> long pause in a serious conversation could be brought to an end by
> one of his thoughtful, cautious, well-phrased constructive comments.
> Equally, it could be terminated by a grunt which, no matter how
> charming the smile that came with it, might seem to the recipient to
> take him nowhere.

Rodger also thought that Myers failed to provide real intellectual
insight into the field of occupational psychology. 'It was not easy to
discern any pattern in his research and teaching programmes. I believe
that this created an important obstacle in the way of the institute's
progress.'[6]

Certainly things got much tougher in the 1930s. The depression
caused fee income to drop away, there was increasing competition
from management consultants and money had to be raised if the
institute was to survive. Fund-raising was something that Myers did
reluctantly and unhappily; it 'became clear that he was not well suited
to the job of raising money, or arguing about it'. He gave public talks
about road accidents or 'the problem of domestic service' and
persuaded members of the royal family to come along and be voca-
tionally tested, but he was not a natural populist; nor, any longer, at
the scientific cutting edge. Myers came under increasing pressure from
the businessmen on his committee who felt he was not commercially-
minded enough.

In 1936 the day-to-day running of the institute passed to a businessman, Clifford Frisby, and Myers was kicked upstairs. He began
spending much of his time in the country. Soon after moving to
London in 1922, he had bought land in the Somerset village of Winford,
a few miles south-east of Minehead, on the edge of Exmoor, and built
a large house with elements both of Scottish baronial and imperial

German, surrounded by extensive grounds which his wife Edith planted out with trees and shrubs. Though an outsider, Myers took an active part in the community. He was president of the local branch of the British Legion, for some years entertained the ex-servicemen of the neighbourhood to dinner each year, and chaired the committee promoting a scheme for a village hall. Although Myers 'did not follow hounds', the local newspaper reported, 'he had given testimony to the Devon and Somerset Staghounds of his goodwill towards the sport by walking puppies and in other ways'. His son, Edmund, however, engaged fully with the life of the country gentleman. He became a passionate huntsman and a fine rider and entered the army in 1926, at the age of twenty.[7]

In July 1939, after he had retired from the National Institute for Industrial Psychology, Charles Myers heard that, with another war looking ever more likely, an official review into shell shock in the Great War was to be carried out. Myers was forced to confront his own role in the process – something he had not been willing to do two decades earlier.

This chapter in Myers' life had begun in April 1920, when an obscure peer, Lord Southborough, had stood up in the House of Lords and suggested that an inquiry into shell shock should be set up. He argued that, because of the level of interest in this subject, it was necessary to learn the lessons of the whole episode in order to be able to apply them in any future war.

A retired civil servant with a reputation as a backstairs fixer, Southborough had almost certainly been set up by the War Office, which had been under considerable public pressure on the issue of military executions and shell shock since the Armistice. The writer A. P. Herbert had published a fictionalised account of one wartime execution, and a Labour MP had repeatedly pressed for an inquiry into the issue. An internal War Office inquiry in 1920, which had ruled that there had been no miscarriages of justice but provided no evidence, had been widely dismissed as a whitewash. At the same time, the issue of pensions for shell-shocked soldiers posed a considerable problem for the cash-strapped government.[8]

Later in 1920, Sir Bernard Cubitt, a senior War Office official, began to organise the committee. Lord Southborough himself was in the chair and various military and medical figures sat on it, three of whom, Drs Maurice Craig, Frederick Mott and William Aldren Turner, had played prominent parts in the shell shock saga.[9] On 7 September 1920 they held their first meeting and decided to invite prominent military and medical witnesses to appear before the committee and to send out a preliminary questionnaire. Following criticism in *The Times* from Dr Charles Wilson, who had served as regimental medical officer with the Middlesex Regiment in the war (and would later, as Lord Moran, be Churchill's doctor and write *The Anatomy of Courage*), it soon cast its net wider and took evidence from a number of front-line doctors and soldiers. It even interviewed some shell-shocked pensioners.

The committee heard from William Rivers, Henry Head, William Brown, Gordon Holmes and many others. It was also very keen to talk to Charles Myers and asked him to give evidence in January. Myers found himself in a difficult position. He said later that to have appeared before the committee (or helped to write the official history) would have been 'too painful'. Perhaps his pride was wounded by the fact that he had not been appointed to sit on the committee himself. But the real reason was different. Myers feared that he, as the man who had coined the term 'shell shock', was being lined up as scapegoat; the last thing he needed, at a time when he was trying to get his new institute off the ground, was damage to his reputation. On 11 January 1921 he sent a bald reply: 'I would prefer to be excused from giving evidence before the War Office Committee of Inquiry into shell shock.' The members were informed that 'Dr C. S. Myers, who would have been a valuable witness, has declined.' The committee could not force him to attend but at its thirtieth meeting on 22 September 1921 it decided to try again. As 'his evidence as neurologist in the early days in France would be of great value to the committee, it is hoped that he may be persuaded to come forward'. But Myers remained obdurate.[10]

For the next decade and a half, Myers had nothing to do with war veterans or their problems. But then in 1939 he heard that the Ministry of Pensions had summoned a group of experts to review experience in 1914–18 and draw up guidelines for official policy. The ministry was worried that millions of British civilians might break down when bombed by German aeroplanes in another war and wanted to know what could be done to prevent it. It was also determined to avoid any repetition of the huge pensions bill for 'shell shock', which had followed the previous war. Nearly all the surviving shell shock doctors, including two veterans of the 1922 Committee and Myers' old dining companion in Boulogne (and later enemy) Gordon Holmes, were roped in.

Myers was upset not to be involved this time and, together with three other shell shock doctors, he submitted a long and unsolicited memorandum to the committee, setting out the line they thought ought to be followed. In fact, there was little difference between the official view and that taken by Myers' more 'psychologically' minded group. Both sides agreed that in a future war quasi-medical terms such as 'shell shock' should be avoided, pensions should be kept to minimum and psychotherapy confined to simple rest and reassurance. The official committee duly concluded that, as *The Times* put it, shell shock had been 'a gross and costly misnomer and should be eliminated entirely from our nomenclature'.[11]

This episode prompted Myers finally to write up his time in France – to issue the report on shell shock he had been writing in 1916. In *Shell Shock in France 1914–18*, published in 1940, he tried to combine two contrary purposes: to defend his own record and to enlighten the public on the nature of the problem. As a result, he produced a muddled book. On the one hand, Myers catalogued all the humiliations and obstructions he had suffered while working in France, his efforts to save men facing the death penalty, and his struggles to get proper facilities for the treatment of 'shell shock'. But at the same time, he had to admit that 'shell shock' was 'a singularly ill-chosen term' and had 'proved a singularly harmful one'. The book's take-home message was tough-minded. Myers explained in the preface that he hoped to convince the general public how dependent 'shell shock' was on 'a previous psycho-neurotic history and inherited predisposition, on inadequate examination and selection of soldiers fitted for the front line, and on lack of proper discipline and *esprit de corps*; and how necessary it may be to adopt apparently harsh measures in order to diminish the undoubted "contagiousness" and the needless prolongation of the complaint'.[12]

His advice was not at first heeded. For the first two years of the Second World War, the British Army resolutely ignored the lessons of the previous conflict and refused to introduce psychological testing of recruits. It only relented in 1941, after the disaster of Dunkirk had revealed what happened when unsuitable men were let into the military and a highly intelligent gunner, Sir Ronald Adam, had become Adjutant General. General Adam grasped at once that simple intelligence testing was not only necessary but would also enable the army to look more modern and less blimpish. Charles Myers was one of those who were brought in to advise on how to implement the scheme and, to his great satisfaction, the staff of the National Institute for Industrial Psychology provided most of the trained personnel to conduct it.[13]

On 8 October 1946, a luncheon was held at the Mansion House in the City of London to celebrate the twenty-fifth anniversary of the foundation of the National Institute for Industrial Psychology. The lord mayor presided, some 230 guests attended and the guest of honour, Sir Stafford Cripps, President of the Board of Trade in the new Labour government, praised the institute's role in wartime and invited its help in meeting the challenges of peace. Industrial psychology, said Cripps, 'once regarded as the hobby of the pioneer, has now become the very core of our production problem'.

Responding to the toast, Myers acknowledged that the institute had had its lean periods. Nonetheless, overall, he was proud of what it had achieved and had no regrets about leaving 'a fairly peaceful academic life in Cambridge in *pure* psychology for a wider, much less tranquil life in *applied* psychology in London'. Myers then told the audience that forty-eight years earlier he had been a member of the Cambridge expedition to the Torres Straits and Sarawak:

> I recall my efforts to induce an old savage named Ulai to part with his tally of love conquests, a bundle of sticks in which each such episode was scored by a notch cut in them. I wanted to bring this tally – *Kupe* was its native name – back to England as a museum piece, but old Ulai would part with it only grudgingly. As at last he handed it over, he gazed at it with a melancholy expression, saying, 'Me old man now.' So I am inclined to feel today, although in fact I am far from 'parting' with the institute. The younger among you may tend to regard me rather as qualifying now for a 'museum piece'.[14]

After the lunch, Myers returned to his house in Somerset. Four days later he died peacefully.

Conclusion

A remarkable event took place in the University of London in the summer of 1911: over two thousand people, from more than fifty countries, gathered together to discuss the question of race and racial differences. The First Universal Races Congress, organised by Gustav Spiller, a Hungarian-born writer on ethical and social questions, and Felix Adler, a German-American professor of ethics, somehow managed to unite liberal supporters of colonialism and left-wing anti-imperialists in an affirmation of common humanity.[1]

Alfred Haddon, the organiser of the 1898 Cambridge expedition to the Torres Straits, hailed the meeting as 'a new departure in the history of the world'. 'During the week of the congress,' he wrote in *Science*, 'there could be seen in the halls of the University of London men and women of all shades of colour and of different religions in friendly converse or planning schemes for breaking down racial and other prejudice, as well as for the betterment of mankind.' In a paper on 'the permanence of racial mental differences', Charles Myers told the conference that the 'mental characters' of the peasant class in Europe were 'essentially the same as those of primitive communities'. Any differences were owing to environment and individual variability. The Cambridge expedition's work with Torres Straits Islanders had, he said, shown that in acuteness of vision, hearing, smell, etc 'these peoples are not noticeably different from our own': 'In temperament we meet just the same variations in primitive as in civilised communities. In every primitive society is to be found the flighty, the staid, the energetic, the indolent, the cheerful, the morose, the even-, the hot-tempered, the unthinking, the philosophic individual.'

This was a bold statement of common humanity. But Myers went further when he came to write up his paper. Having in the meantime read the French anthropologist Lucien Lévy-Bruhl's book, *Les Fonctions mentales dans les sociétés inférieures* (later translated as *How Natives Think*), which argued that the primitive mind was 'mystic' and 'pre-logical' and incapable of rational thought, he now felt it important to rebut that view. On the contrary, declared Myers, 'There is not a savage who cannot talk logically about matters of everyday life. He *can* reason as we do. He *will* not, where the force of social tradition is so strong.' Here Myers was drawing not only on his psychological tests but on his direct human contact with helpers and houseboys on Murray Island.[2]

Something important had happened. A decade after the expedition's return, its findings – originally so tentatively expressed – were now being vigorously talked up; used, indeed, to fuel a political debate about race. It marked the beginning of an intellectual shift – one which would slowly transform 'race' from a biological to a cultural concept and ultimately dismantle it as a tool of science altogether – until finally, in 1950, UNESCO, the newly formed international cultural organisation, issued a declaration stating that there was no scientific basis to the idea that human groups differed in 'their innate capacity for intellectual and emotional development' and that there was 'no reliable evidence that disadvantageous effects are produced' by the crossing of races.[3] How far that process was driven by changes in scientific evidence and how far by ideology – and, in particular, by a liberal revulsion at the racial views of the Nazis – is a matter of debate for historians. The historical significance of the 1898 Cambridge expedition was that it helped to set this ball rolling.

However, that statement must immediately be qualified. This is not a simple story, but one beset by paradoxes, surprises and wrong turnings. For a start, even as Myers was trumpeting the expedition's findings, the means by which they had been arrived at were increasingly being called into question. Although Haddon's expedition had won (and would always retain) a place in the history of anthropology, as the first attempt to do scientific fieldwork among primitive people, no one seemed to be in any hurry to emulate its method – apart from the American psychologist R. S. Woodworth, who carried out an inconclusive set of tests on visitors of different races at the 1904 St Louis Fair. The expedition members themselves did not persevere in this area: although Haddon, Rivers, Myers and Seligman all sat on the committee which produced the fourth edition of *Notes and Queries on Anthropology*, the Royal Anthropological Institute's bible for the profession, and wrote many of the entries, the publication which emerged in 1912 did not mention the psychological testing of primitive people at all. Instead, it contained a lengthy explanation of the 'genealogical method' by Rivers.[4]

Four years later the professional *coup de grâce* was administered by the American psychologist, Edward Bradford Titchener, a pupil of Wilhelm Wundt, the father of experimental psychology. In a lengthy review of the Torres Straits reports, Titchener expressed 'a complete mistrust' of McDougall's 'formal procedure', and found that his conclusions did not follow from his experimental data and his method was 'full of suggestion'. By contrast, he acknowledged Rivers' work on colour vision to be a very painstaking piece of research which stated its conclusion with becoming caution – 'that the colour vision of the Papuan is characterised by a certain degree of insensitiveness to blue (and probably green) as compared to that of Europeans'– but he doubted 'whether the observations made warran[ted] the inference drawn from them'. Titchener raised a host of other objections, such as the fact that insufficient allowance had been made for the difference between testing colour vision in tropical sunshine and in a school room in Northern Europe.[5]

Moreover, if we look more closely at what Haddon and Myers said at the 1911 Universal Congress of Races, they were not as liberal-minded as they first appear. Haddon's report in *Science* described the occasion as 'pathetic as well as inspiriting'. Sentiment and rhetoric might initiate reforms, he wrote, but it would take years of hard practical work to make racial equality happen. He cautioned against going too fast. Similarly, although Myers argued that the differences between the white and negro races were ultimately due to environment – and could therefore be eliminated by a change of environment – he thought such a process would take 'many hundreds of thousands of years'.[6]

How then could the UNESCO committee announce, less than forty years later, that the concept of race was scientifically worthless? The process by which the 'evolutionist caveats' voiced by Haddon and Myers came to be discarded was too long and complicated to be retold in detail here. The fundamental point is that the main players in changing the significance of race in science were British biologists and American anthropologists and psychologists – and Hitler. After the First World War a new generation of mathematically trained British geneticists teased out the implication of Mendelian chromosome theory and showed that the interaction between nature and nurture in both animals and human beings was much more complex than the eugenicists, with their 'stud book mentality', had suggested. In a parallel development, a decade earlier, the German-American anthropologist Franz Boas had carried out a study of the skull shapes of recent immigrants to the United States and found that once in America head shapes tended to conform – a conclusion which did not win general acceptance among scientists but helped to undermine the status of skull measurement as an indicator of racial type and of physical anthropology in general. Finally, in the 1920s and 30s re-analysis of the mental tests carried out on recruits to the US Army in the war showed that, hidden within the data generated, was a contradictory fact – that black recruits educated in the Northern states scored better than whites educated in the South. Since the Army tests had been used to establish the apparently inferior intelligence of black people as a scientific fact this was a very telling critique.[7]

What place, then, do Rivers, McDougall, Myers and Elliot Smith occupy in this narrative? Rivers died before the important discoveries had entered the scientific mainstream. Myers was clearly sympathetic to the more egalitarian atmosphere and by 1932 had considerably shortened his view of the time-span needed for non-whites to 'catch up'; but his position at the National Institute prevented him from taking a public stance. McDougall, after the furore caused by his lectures in America in 1920, kept off the subject of race, though his sympathy with the views of his neighbours in North Carolina was obvious.[8]

Elliot Smith's position was more complex. In his book on shell shock he had argued powerfully that environment, not heredity, could cause mental illness. His diffusionism and the pessimistic view he took of human history also swayed him in different directions, as was apparent in the Galton lecture he gave to the Eugenics Society in 1924 where he stated that 'different races have obvious differences in mental aptitude', yet also asserted that 'there is no necessary connection between race and culture'. After the Nazis' arrival in power, however, Elliot Smith was active in opposing their racial doctrines in various intellectual forums. Despite suffering a stroke in 1932, he worked hard to persuade the Royal Anthropological Institute to come down against the Nazis' theories – circulating copies of Franz Boas' pamphlet 'Aryans and non-Aryans' to committee members – but though he was now much enfeebled he remained a divisive rather than a unifying force. The Institute's Report on Race and Culture which finally appeared in April 1936 was a weak compromise that satisfied no one. Elsewhere, however, Elliott Smith made his views well known, using his position as Chairman of the International Congress of Anthropology and Ethnology in London in 1934 to deliver a robust rebuttal of 'Nordic race claims'. A year later, an article of his on 'the Aryan question' concluded that it was 'certain that there is no race living at the present time which can truly be called "Aryan"'.[9]

The most publicly visible figure of all, however, was Alfred Haddon, by now in his late seventies. Together with the biologist Julian Huxley, Haddon was the author of *We Europeans*, published by Jonathan Cape in 1935, a book designed to make the scientific arguments against Nazi racial theories accessible to a general readership. Drawing on recent biological and anthropological work, Huxley and Haddon argued that the scientific theory of race, as applied to human beings, had 'lost any sharpness of meaning. Today it is hardly definable in scientific terms, except as an abstract concept ... in the ultimate analysis the matter must be incapable of scientific determination since the decision as to what is a "race" is a personal matter resting largely on subjective impressions'.[10]

We Europeans was not quite what it seemed, however. It was essentially the brain-child of Charles Seligman and the medical historian Charles Singer, both Jewish and much involved in helping Jewish refugees from Germany, who felt that the book would command more authority if authored by gentiles. Julian Huxley, the grandson of Haddon's patron, Thomas Henry Huxley, had held somewhat different views only a decade earlier – there was, he wrote in *Africa View*, published in 1931, 'a certain amount of evidence that the negro is an earlier product of human evolution than the Mongolian or the European, and as such might be expected to have advanced less, both in body and mind'. For his part, Haddon seems to have had mixed feelings about his role, believing that he had been 'powerless to alter the rather bitter controversial trend'. 'I have from the beginning been very sorry to have had anything to do with it', he complained privately, yet to Huxley he wrote, 'I really am pleased with the book – it is much needed – no more than in Germany – where it will be banned.'[11]

When Haddon died, in his eighty-fifth year, in Cambridge on 20 April 1940, the Nazis were poised to invade Western Europe. A decade later, when UNESCO issued its declaration on race, the political and cultural landscape had been completely transformed. Yet the UNESCO declaration did not end discussion on race within science – indeed it provoked an immediate outcry from physical anthropologists – and there would be many more years of argument about race and intelligence among psychologists. Moreover, it would take decades before the views expressed by UNESCO were reflected in changing public attitudes. And scientifically speaking, the UNESCO declaration represented something of a high-water mark in environmentalism. Almost immediately, the pendulum began to swing back towards heredity. The years since Crick and Watson discovered the secret of DNA have seen a renewed emphasis on genetics and on race, now rebranded as 'diversity'. Thus, at the time of writing, Oxford University is carrying on an elaborate survey of the British population, mapping its DNA in order to establish 'what really happened' when the Anglo-Saxons settled in Britain. Although its methodology is much more sophisticated, the project is strikingly similar to the abortive anthropometric survey of the British Isles which Alfred Haddon proposed in the 1890s – and there is the same emphasis on isolated communities which have remained racially homogeneous, though the word 'race' is of course not used.[12]

When William Rivers died suddenly in 1922 his reputation stood at its highest. His friends Henry Head, Alfred Haddon, Charles Myers, Charles Seligman and Grafton Elliot Smith each produced biographical tributes emphasising different aspects of his many-sided career. It seemed that a battle for his legacy was about to begin.[13]

Yet within a few years the whole edifice – Rivers' work and his reputation – had begun to crumble and his claim to have made ethnology a science was being called into question. It wasn't simply that Bronisław Malinowski, the dominant figure in British anthropology from 1925 on, rejected Rivers' entire agenda; Rivers' own students, with an oedipal instinct to which British anthropologists seem particularly prone, also found flaws in his work. Malinowski repudiated the 'gene-alogical method' and regarded 'kinship' as a 'spuriously scientific and stilted mathematization': both, to him, were examples of the 'false problems' which had entered anthropology. 'Another false problem', he wrote in 1930, 'is that of the origin and significance of classificatory systems of nomenclature.' Instead, he argued, it was best to put aside 'kinship algebra' for 'full-blooded sociological research'.[14]

Meanwhile, Rivers' former pupils at Cambridge were revisiting his work on Melanesia. Rivers had been mystified by the anomalous marriage forms there and had, after long cogitation, decided that there was a gerontocracy, a society in which old men married their grand-daughters. In 1926 a brilliant young student called Bernard Deacon carried out several months of research on the island of Ambrym in the New Hebrides which Rivers had visited in 1914 (after abandoning John Layard on Atchin). Deacon found serious methodological flaws in an unpublished paper Rivers had left after his death but died himself before he could publish his research. It was left to Rivers' pupil, A. R. Radcliffe-Brown, to demolish his whole explanation of marriage arrangements in Melanesia. Class was the answer, he argued, not men who married their granddaughters. The theory of a Melanesian geron-tocracy proposed by Rivers simply did not stand up. The historical reconstruction of Melanesian society – being based on an interpret-ation of these arrangements – was also rendered worthless. Radcliffe-Brown differed from Malinowski in regarding kinship as important, but agreed with him that 'the functional study of society' was far more important than 'conjectural history'. Thus, as Adam Kuper has written, 'within five years of Rivers' death, and thirteen years after the publication of his *History of Melanesian Society*, his outstanding student had turned upside down the conjectural history of Melanesia which had been his monument.' Rivers' book came to be widely regarded as 'a complete waste of a good brain's time'.[15]

Another important pillar of Rivers' intellectual world, the 'nerve experiment' he had conducted with Henry Head, suffered a similar fate. Something about the published accounts aroused the scepticism of the surgeon Wilfred Trotter, who embarked on an elaborate attempt to duplicate the experiments. Using more thorough methods, he reported very different results. In 1921, Francis Walshe, a colleague of Trotter at University College Hospital, wrote a more detailed critique and showed it to Trotter. By then Head was developing Parkinson's disease and Trotter persuaded Walshe not to publish while the great man was alive. Consequently Walshe's paper did not appear until 1942, two years after Head's death. In what has been described as a 'comprehensive annihilation', Walshe argued that Head had made no effort to establish whether the skin actually contained the anatomical structures needed to support his idea of a split between 'protopathic' and 'epicritic' nervous systems. His theory, Walshe wrote, 'succumbs to the danger that always besets abstract thinking: that of confusing thoughts with things'.[16]

Rivers, as we have seen, had used the concept of 'protopathic' and 'epicritic' levels in the nervous system to underpin his final account of shell shock. Perhaps for that reason, his theories in this field had little enduring influence. A review of the shell shock literature published in 1940 by a group of psychologically oriented British doctors referred briefly to Rivers' work but in the war that followed it was scarcely mentioned: the main emphasis was on new methods of treatment using barbiturate drugs. And when medicine returned to the 'war neuroses' in the 1970s, after years of neglect, the Americans who now drove the debate ignored Rivers completely. Nor is Rivers regarded as among the fathers of Post-Traumatic Stress Disorder, the term invented by the American Psychiatric Association in 1980.[17]

There were always two sides to Rivers' reputation: the work and the man. In his earliest tributes, Frederic Bartlett kept the two separate, voicing gentle reservations about the work while celebrating the man. Bartlett, who took over from Myers at Cambridge in 1922 and was created a professor a decade later, turned the Cambridge psychology department into the most respected in the country, while carefully keeping within the experimental constraints Myers had chafed against. Best known for his 1932 book *Remembering*, Bartlett was himself a great rememberer who did much to keep the memory of Rivers alive in his university. It is largely thanks to Bartlett that, as the Cambridge historian of science, Simon Schaffer, has put it, 'Rivers is a ghost who still haunts this town, still fascinates and provokes thought.' This ghost is very much 'early Rivers', 'Rivers before Elliot Smith'.[18]

But the most important reason why Rivers' memory was kept alive was that a great writer chose to commemorate him. In 1926 Siegfried

Sassoon began reworking his life in lightly fictionalised memoirs. *Memoirs of a Fox-Hunting Man* (1928) dealt with the pre-war years, *Memoirs of an Infantry Officer* (1930) with the trenches, and finally, in 1936, in *Sherston's Progress*, Sassoon reached his time at Craiglockhart. Although he had been forced in the earlier volumes to fictionalise a good deal, Sassoon could name Rivers, now long dead, and write about him directly. He created a loving and detailed portrait of his friend and mentor: 'Much as [Rivers] disliked sending me back to the trenches, he realized that it was my only way out. And the longer I live the more right I know him to have been.'[19]

'We simply love your book', Rivers' sister, Katherine, wrote to Sassoon soon after publication in September 1936. 'So glad you called him "Rivers". Now his name "liveth forever more" on earth.' She was proved right. Over the next few decades, much of the Great War literature fell into neglect, but a core of memoirs and poems – by Wilfred Owen, Robert Graves and Edmund Blunden, as well as by Sassoon – remained in print and defined the war for new generations, despite the protests of military historians that this 'literature of disenchantment' was quite untypical of British experience in the trenches. This literary tradition was revitalised in the 1970s by Paul Fusssell's masterly *The Great War and Modern Memory*, which familiarised another generation with Sassoon, Rivers, and shell shock. A decade later – in a climate defined by the Vietnam War and the feminist movement – the literary critic, Elaine Showalter, included a chapter on Rivers and 'male hysteria' in her influential book *The Feminine Malady*, making a rhetorical contrast between Rivers' use of the 'talking-cure' and the 'persuasion backed up by electricity' employed by Dr Lewis Yealland at Queen Square, and thus providing Pat Barker with one of the foundations of her *Regeneration* trilogy.[20]

Although Barker's trilogy was not universally praised, it sold enormously and the third volume won the Booker Prize. It repackaged Rivers for a more literal-minded generation and recruited him into the psychotherapy movement of the 1980s. Now women had heard of him. Psychotherapists held conferences devoted to him. He became the patron saint of psychotherapy.[21]

Charles Myers' legacy lay more in institutions than in ideas: as Bartlett said, 'he built a laboratory, a society, an institute'. When the main institution he created, the National Institute for Industrial Psychology, closed its doors in 1971, after limping on for two decades following his death, that legacy was ended. Yet, in many ways the philosophy of rational-minded, harmonious industrial relations Myers stood for has been belatedly adopted. With the end of the class war and the collapse of most of traditional manufacturing industry, British management has turned to the Japanese model – and in the process embraced much that Myers was preaching.

Most of Myers' psychological writings are long forgotten but his wartime work lives on principally because the method of treating shell-shocked soldiers which he advocated in 1916 has become the basis of modern military psychiatry. However, Myers' claim to be the father of this method is a tangled one. As we have seen, he got the idea of treating soldiers near the front line from the French. It was then taken up by the American psychiatrist Thomas W. Salmon when he came to Europe and talked to British and French doctors, and became known in the United States as the 'Salmon plan'. During the Second World War, the US Army did not at first feel it necessary to have psychiatrists near the front line but, after sustaining heavy psychiatric casualties in the Tunisian campaign in 1942–3, it reluctantly accepted the need to bring them in.

In the 1950s, the 'Salmon Plan' became known by the acronym PIE – which stood for proximity, immediacy, and expectancy. The soldier was to be treated near the front line (proximity), as quickly as possible (immediacy), and in a military atmosphere which would encourage him to return to his unit (expectancy) and would not turn him into an invalid incapable of any further military service. This remains the principal rationale of military psychiatry today and has, with some modifications, formed the rubric for what has been done with British and US forces in Iraq and Afghanistan. However, there have always been sceptics about forward psychiatry. We have seen that during the First World War there were no firm statistics about the rate of relapsing and it was widely believed that such soldiers were 'quite useless because no one would trust them in the line'. A recent historical review has doubted whether 'forward psychiatry' has actually been as effective in getting soldiers back to fighting or in preventing them becoming psychiatric cases later on as its advocates would have us believe, a view which has itself been challenged by the military authorities.[22]

The eclipse in the reputation of Grafton Elliot Smith was not as immediate as might have been expected. This was principally because in the decade after his death, the doctrine of diffusion in prehistory was taken up by his compatriot, Vere Gordon Childe, who became the most prominent and articulate figure in British archaeology in the middle of the twentieth century. Childe was strongly influenced by Elliot Smith and in his best-selling Penguin paperback *What Happened in History*, published in 1942, offered a diffusionist account of how the barbarism of Europe had been irradiated by the civilisation of the East, from which most of the important inventions had come. In his autobiography Childe wrote that 'The sea-voyagers who diffused culture to Britain and Denmark in the first chapters of the first Dawn ... though they do not hail from Egypt, yet wear recognisably the emblems of the Children of the Sun.'[23]

After the Second World War, however, came three important developments that completely transformed our understanding of prehistory and made Elliot Smith's theories seem as remote as early Christian theology: Raymond Dart's *Australopithecus africanus* was pronounced to be genuine; radiocarbon dating techniques were developed; and Piltdown man was revealed to be a fraud.

As we have seen, Elliot Smith and Arthur Keith refused to accept as a human ancestor the *Australopithecus africanus* fossil which Raymond Dart had announced to the world in 1925, in part because they had not been able to examine the fossil personally. However, in 1947 the British anatomist Wilfred Le Gros Clark went to South Africa to see it for himself and, after examining Dart's fossil skull of a child and other similar finds which had by then been made nearby, proclaimed, with the high gravitas of an Oxford professor, that *Australopithecus* was in fact a human ancestor, a hominid. 'African genesis' – the idea that the beginnings of humanity lay in Africa – now became the dominant narrative of prehistory, and was reinforced from the 1960s on when Richard Leakey and other scientists found further, more overtly human, fossils in East Africa.[24]

Elliot Smith's refusal to accept *Australopithecus* had of course been only half the story; the other being his acceptance of Piltdown Man.

Here too there came post-war revelations. In 1953, longstanding doubts about Piltdown's authenticity came together with new methods of dating: it was conclusively established that Piltdown was a forgery – a crude combination of a human skull with an ape's jaw – though the identity of the forger was (and has remained) disputed. Here, too, new evidence showed that Elliot Smith was completely wrong.[25]

The most momentous change, however, was the arrival of the new technology of radiocarbon dating, which emerged as by-product of the Hiroshima bomb. When finally refined and applied, it showed decisively that the megaliths of Western Europe, such as Stonehenge, were over 1000 years older than the pyramids of Egypt, which dealt a fatal blow to Elliot Smith's idea that culture had been diffused from ancient Egypt. At a conference marking the centenary of his birth, in London in 1972, Elliot Smith's reputation in anatomy and comparative neurology stood up reasonably well, but the archaeologists and anthropologists queued up to kick him: 'What he taught us as regards ethnology was absolute rubbish', the anthropologist Edmund Leach declared. The real question, for most of the speakers, was not whether culture was diffused from Egypt but 'what went wrong with Elliot Smith, the brilliant anatomist whose work is still revered in the anatomical field at Sydney?'[26]

By now, the arch-prophet of diffusion had become a figure of fun, a tethered goat to be slaughtered in the early chapters of books on the radiocarbon revolution. Elliot Smith's rich, overripe prose lent itself easily to mockery. Yet some authors went so far in reacting against Elliot Smith as to deny the importance of *any* diffusion, and almost returned to the nineteenth-century idea of cultures existing in isolation. That approach has in turn come under attack since the 1990s as yet another new technology has come to the fore in archaeology – genetic testing, using a population's DNA to ascertain its past. Although some researchers believe that we now place too much reliance on genetics, this movement has generally swept all before it and redefined the agenda of archaeology. The interesting thing about the modern synthesis is that it re-emphasises the role of diffusion and of race (detoxified as 'diversity') in human evolution. And some of its discoveries are vindicating Rivers and, even, Elliot Smith's hunches – though of course the modern chronology is quite different from that of the 1920s. Rivers' interpretation of Melanesian history – as consisting of successive waves of invaders – was more or less right, according to a genetic study carried out by Temple University in 2008. Elliot Smith was right to assign to the Phoenicians a leading role in the spread of culture and there may even be some elements derived from Egypt in South American culture. We now know that the mummy from Torres Straits that Elliot Smith examined in 1914 was only about thirty years old; yet the conclusion he leapt to – that the people living in the Torres Straits had come from Africa many centuries before – has been confirmed by genetic testing. We now have evidence that, 50,000 years ago, 'beachcombers' from the Horn of Africa made their way along the coastline of Asia until they reached Australasia.[27]

In the preface to the twenty-third edition of his *Introduction to Social Psychology*, written in October 1936, William McDougall claimed that the various sects of psychology at the time – 'the Psycho-analysts, the Gestaltists, the Behaviourists, the Connectionists, the Characterologists, the Social Psychologists of America, the cautious middle-of-the road men' – had all moved further towards the theory of human behaviour as being ruled by instincts which he had set out in his book in 1908. He then predicted that, after a lapse of some few years when his name would be completely forgotten, these principles would be 'generally accepted as the main pillars of a psychology which will serve as the indispensable basis of all the social sciences'.[28]

McDougall was half right. Many of the intellectual fashions of his day (even behaviourism) have run their course and psychoanalysis, which reigned triumphant in American psychiatry for half a century after the Second World War, is now in complete decline. However, there has been no great rediscovery of McDougall's work.

He had chosen a bad time to die. Within less than a year of his death in November 1938, the war had brought very different preoccupations. Although his son, Kenneth, dutifully continued the Lamarckian experiments with rats for a short while, he had abandoned them well before he joined the US Army in 1943. When Kenneth was accidentally shot in France in September 1944 by a member of his own unit – an 'edgy fellow' who had been shell-shocked at Anzio and had only just rejoined the platoon – a chapter closed. The only people who kept McDougall's flame burning were the psychic researchers.[29]

Today, McDougall's position is complex. On the one hand, his name is invoked by those psychologists who argue that there is a connection

between race and intelligence and execrated by those who argue that
there is not. Yet, at the same time, he is recognized as one of the
founders of ethology, the science of animal behaviour, and an important influence on both Konrad Lorenz and Niko Tinbergen. McDougall's
most important British pupil, Sir Cyril Burt, was exposed after his death
as a fraud who falsified the results of his research, then partially re-
habilitated. His last American pupil, Jerome Bruner, is revered and
honoured and one of the grand old men of American intellectual life.
Margaret Boden, a pioneer of using computer models in psychology,
has recently recalled how she was fired up when she first encountered
McDougall's 'rich theory of the purposive structures underlying normal
and abnormal psychology' in the 1960s. 'What excited me about
McDougall', she wrote, 'was his deep insight into the complex structure
of the human mind and his many explicit arguments against psychological reductionism – most, if not all, I judged to be valid.'[30]

When McDougall began working in psychology the possibility
existed of creating a physiological psychology – of combining the
work of the physiological researchers with the insights into the mind
which the philosophers and psychologists had developed. Why did
McDougall not pursue this idea? Why did he succumb to animism
and abandon his early work? Because, some have suggested, he was
inhibited by the criticism of philosophers. McDougall himself said it
was in order to reach a wider public. But there was another reason:
McDougall realised that physiology could never supply all the answers.

Although many of the answers which Rivers, McDougall and Elliot
Smith gave have proved to be wrong, because in their time the data
to answer them did not exist, the questions they posed are still
relevant. On the face of it we now live in a completely new world.
All the old gods are dead – neither nationalism, Marxism, psychoanalysis nor Christianity any longer provide philosophical ballast.
Instead, the modern intellectual landscape is dominated by two
phenomena, Neo-Darwinian genetics and modern neuroscience – just
as it was in the 1890s.

The achievements of modern neuroscience are enormous. As the
medical polymath Raymond Tallis puts it, 'hardly a day passes
without yet another breathless declaration in the popular press about
the relevance of neuro-scientific findings to everyday life'. It has

colonised old disciplines such as philosophy and psychology, including psychology of the kind taught in Cambridge in the 1890s. And disciplines such as archaeology and anthropology – having over the last century passed through functionalism, structuralism, processualism, post-processualism and Gawd knows what else, and emerged intellectually naked the other side – have now jumped on the neuro-band wagon, in a desperate search for function, fashion and funding. The unity – the 'united biology'– which Elliot Smith strove for has, paradoxically, arrived. Professor Lord Renfrew, no less, now offers a course in 'neuroarchaeology'.[31]

What will this produce? Probably not as much as the hype would have us believe. The obvious and inevitable limitations of neuroscience are often forgotten. Raymond Tallis is sceptical: 'If you come across a new discipline with the prefix "neuro" and it is not to do with the nervous system, switch on your bullshit detector', he has written.[32]

And will Darwinism too prove unable to explain human behaviour, just as it did a hundred years ago? Maybe the archaeologist V. Gordon Childe was right when he declared that 'evolution does not purport to explain the mechanism of cultural change. It is not an account of *why* cultures change – that is the subject matter of history – but of *how* they change.'[33]

Acknowledgements

At the time of the events narrated in this book the stretch of water lying between Australia and Papua New Guinea was called the Torres Straits; today it is known as the Torres Strait. The plural form has been used here.

Extracts from the Haddon Papers and Charles Myers' Torres Straits Journal (Add MS 8073) are used by kind permission of the syndics of Cambridge University Library. I also thank the following libraries for material from their collections: Bethlem Royal Archive (patient case-books); Boston Public Library (papers of Hugo Münsterberg); British Library (Warren R. Dawson papers, Add MS 56303); David M. Rubinstein Rare Book and Manuscript Library, Duke University (William McDougall papers); London School of Economics Library, Special Collections (papers of Charles Seligman and Bronisław Malinowski); the Medical Committee of the National Hospital for Nervous Diseases (hospital archives and consultants' casebooks); Woodward Library, University of British Columbia (letters from William McDougall to Charles Sherrington); University College London Archives (William J. Perry papers); John Rylands Library, Manchester (William McDougall's letters to Samuel Alexander; Grafton Elliot Smith papers); Wellcome Library (papers of Charles Myers, the British Psychological Society, and the National Institute for Industrial Psychology). Pictures have been provided by Duke University (8–10), the Imperial War Museum (13–14), the Museum of Archaeology and Anthropology, Cambridge (1–5) and Pathé (15–16). If any other sources have not been identified, the author and publisher will be happy to rectify the omission at the earliest opportunity.

I am grateful to librarians, archivists, and staff at Bethlem Royal Archive (Colin Gale); Bodleian Library; Boston Public Library

(Kimberly Reynolds); Bristol University Library; Woodward Library, University of British Columbia (Charlotte Beck); British Science Association; British Library; British Psychological Society (Mike Maskell), Cambridge University Library, Department of Manuscripts (Frank Bowles); Cambridge University Medical School Library; Cambridge University Museum of Archaeology and Anthropology (Wendy Brown); the David M. Rubinstein Rare Book and Manuscript Library, Duke University (David Pavelic and Rachel Ingold); King's College, Cambridge (Tracy Wilkinson); London Library; London School of Economics Library (Special Collections), John Rylands Library, Manchester; McMaster University Library (Bev Beyzat); Imperial War Museum, Department of Documents; National Archives, Kew; the National Hospital for Nervous Diseases (Jackie Cheshire); Selwyn College, Cambridge; St John's College Cambridge (Malcolm Underwood and Kathryn McKee); University College London Archives; Yale University Library (Kristen McDonald).

A travel grant from the Wellcome Trust enabled me to consult the McDougall papers at Duke University; my thanks to David Clayton for helping me through the hoops. Between 2004 and 2012, I was a Visiting Research Fellow in the Changing Character of War Programme at Oxford, which was funded first by the Leverhulme Trust and then by the University of Oxford. I owe this honour – and the many stimulating contacts with military historians it provided – to Hew Strachan.

My intellectual debts to the work of John Forrester, Anita Herle, Henrika Kuklick, Adam Kuper, Ian Langham, Graham Richards and George W. Stocking, Jr., will be apparent.

I am grateful, in all sorts of ways, to Sophie Baker, Peter Barham, Helen Bettinson, Margaret Boden, David Cohen, John Dollar, Brian Durrans, John Forrester, the late Richard Gregory, Liz Hartford, Rhodri Hayward, the late Leslie Hearnshaw, Anita Herle, Boyd Hilton, Barrie Houghton, Marylla Hunt, Tim Jeal, Edgar Jones, Lynne Jones, Chris Keil, Henrika Kuklick, Ute Leonards, Ruth Leys, Richard McNally, Charlotte Moore, the late Brigadier Edmund Myers, Matthew Parker, Jessica Reinisch, Graham Richards, the late Joan Rumens, Loesje Sanders, Nick Scott-Samuel, Simon Wessely, Allan Young and Elizabeth Whipp.

I particularly wish to thank my agent, Clare Alexander, for her enthusiasm and help. This book was commissioned by Jörg Hensgen, and the text has benefited enormously from his skill and judgement: he is a prince of editors. Will Sulkin was also involved in the commission; it was a privilege to be published by him. Will's successor at the Bodley Head, Stuart Williams, and his team have been hugely supportive. My stepbrother Prosper Devas created the maps – and cheered me up. My sister, Caroline Moser, a wonderful supporter of my work, once again gave me a roof in London. As always, though, my greatest debt is to my wife, Sue, who has lived with this subject for many years. She, more than anyone, got me to the finishing post.

Abbreviations

Butler	A. G. Butler, 'Moral and mental disorders in the war of 1914–18', in *The Australian Medical Services in the War of 1914–1918*, Vol. 3 (Canberra, 1943)
CGS/TSD	Charles Gabriel Seligman, 'Torres Straits Diary', Seligman papers, London School of Economics
CSM/CSS, 1–5	Charles S. Myers, 'Contributions to the study of shell shock' 1. 'Three cases of loss of memory, vision, smell and taste, admitted to the Duchess of Westminster's war hospital, Le Touquet', *Lancet*, 13 February 1915 2. 'Certain cases treated by hypnosis', *Lancet*, 8 January 1916 3. 'Certain disorders of cutaneous sensibility', *Lancet*, 18 March 1916 4. 'Certain disorders of speech, with special reference to their causation and their relation to malingering', *Lancet*, 9 September 1916 5. 'Unsettled points needing investigation', *Lancet*, 11 January 1919
CSM/*HPA*	'Charles Samuel Myers', in C. Murchison (ed.), *A History of Psychology in Autobiography*, Vol. 3 (Worcester, Mass., 1936)
CSM/Influence	C. S. Myers, 'The influence of the late W. H. R. Rivers', presidential address to the Psychology section of the British Association for the Advancement of Science, 1922. Reprinted in Rivers, *Psychology and Politics*
CSM/*SSIF*	C. S. Myers, *Shell Shock in France 1914–18* (Cambridge, 1940)
CSM/TSD	C. S. Myers, 'Torres Straits Diary', Add MS 8073, Cambridge University Library
CSM/WD	C. S. Myers, 'War Diary', 17 October 1914 to 11 May 1915. Copy in possession of the author
Dawson	Warren R. Dawson (ed.), *Sir Grafton Elliot Smith. A Biographical Record by his Colleagues* (London, 1938)

FCB/CE	F. C. Bartlett, 'Cambridge England, 1887–1937', *American Journal of Psychology* 50 (1937), pp. 97–110
FCB/RDM	F. C. Bartlett, 'Remembering Dr Myers', *Bulletin of the British Psychological Society* 18 (1965), pp. 1–10
GES/Biog	W. R. Dawson, 'A general biography', in Warren R. Dawson (ed.), *Sir Grafton Elliot Smith. A Biographical Record by his Colleagues* (London, 1938)
GES/Fragments	Sir Grafton Elliot Smith, 'Fragments of an autobiography', in Warren R. Dawson (ed.), *Sir Grafton Elliot Smith. A Biographical Record by his Colleagues* (London, 1938)
GES/Papers	Papers of Sir Grafton Elliot Smith, in papers of Warren R. Dawson, Add MS56303, British Library
Head-Hunters	A. C. Haddon, *Head-Hunters: Black, White, and Brown* (London, 1901)
Herle and Rouse	A. Herle and S. Rouse (eds.), *Cambridge and the Torres Straits. Centenary Essays on the 1898 Anthropological Expedition* (Cambridge, 1998)
HP	Haddon papers, University of Cambridge Library
IWM	Imperial War Museum, Department of Documents
JF/EF	J. Forrester, 'The English Freud. W. H. R. Rivers, dreaming and the making of early twentieth-century human sciences', in S. Alexander and B. Taylor (eds.), *History and Psyche: Culture, Psychoanalysis and the Past* (Basingstoke, 2012)
JHBS	*Journal of the History of the Behavioral Sciences*
Jones and Wessely	E. Jones and S. Wessely, *From Shell Shock to PTSD. Military Psychiatry from 1900 to the Gulf War* (Hove, 2005)
Langham	Ian Langham, *The Building of British Social Anthropology. W. H. R. Rivers and his Cambridge Disciples in the Development of Kinship Studies, 1898–1931* (Dordrecht, 1981)
NA	National Archives, Kew
ODNB	*Oxford Dictionary of National Biography*
ONFRS	*Obituary Notices of Fellows of the Royal Society*
RWOCISS	Lord Southborough (chairman), 'Report of the War Office Committee of Inquiry into Shell Shock' (1922)
Slobodin	R. Slobodin, *W. H. R. Rivers* (New York, 1978)
Stocking	G. W. Stocking, Jr., *After Tylor. British Social Anthropology 1888–1951* (Madison, Wisc., 1995)
TSR	*Reports of the Cambridge Anthropological Expedition to the Torres Straits*, 5 vols (Cambridge, 1901–35)
WHRR/*C&D*	W. H. R. Rivers, *Conflict and Dream* (1923)

WHRR/*P&E*   W. H. R. Rivers, *Psychology and Ethnology* (ed. G. Elliot
                Smith) (1926)

WHRR/*P&P*   W. H. R. Rivers, *Psychology and Politics and other essays*
                (London, 1923)

WJP   William J. Perry Papers, University College London

WMcD/*HPA*   'William McDougall', in C. Murchison (ed.), *A History of
                Psychology in Autobiography*, Vol. 1 (Worcester, Mass., 1930)

WMcD/*OAP*   W. McDougall, *An Outline of Abnormal Psychology* (New
                York, 1926)

WMcD/Papers   William McDougall papers, Duke University Library,
                Durham, North Carolina

WoN   B. Shephard, *A War of Nerves. Soldiers and Psychiatrists,
                1914–1994* (2000)

Notes

Unless otherwise specified, London is the place of publication.

Introduction

1 CSM/Influence, p. 8.
2 Eight hundred men from the Newfoundland Regiment took part in the attack; about 110 survived unscathed and of the 700 who fell, 324 were killed. Afterwards, the Divisional Commander was to write of the Newfoundlanders' effort: 'It was a magnificent display of trained and disciplined valour, and its assault failed of success because dead men can advance no further.'
3 L. Hearnshaw, 'Sixty years of psychology', *Bulletin of the British Psychological Society* (January 1962), pp. 2–10.
4 G. Richards, 'Sir Grafton Elliot Smith', *ODNB*.
5 R. Rosenbaum, 'The great Ivy League nude posture photo scandal', *New York Times* magazine, 15 January 1995; J. Marsh, *Project Nim* (documentary, 2011); I. McEwan, 'Beware of arrogance: retire nothing', *Observer*, 12 January 2014.

Chapter One

1 Alfred Haddon, 'Journal' 1898–9; HP 1030; CSM/TSD, 10 March 1898.
2 HP 1030.
3 A. H. Quiggin, *Haddon the Head Hunter* (Cambridge, 1942); H. J. Fleure, *ONFRS* 3 (1939–41), pp. 449–65; A. H. Quiggin and E. S. Fegan, 'Alfred Cort Haddon, 1855–1940', *Man* 40 (1940), pp. 97–100.
4 Quiggin, *Haddon*, pp. 10–11.
5 G. L. Geison, *Michael Foster and the Cambridge School of Physiology. The Scientific Enterprise in Late Victorian Society* (Princeton, 1978).
6 A. Desmond, *Huxley* (London, 1998).
7 Quiggin, *Haddon*, p. 33.
8 Dr Lloyd Prager, quoted in ibid., p. 62.
9 Ibid., pp. 56–8.
10 Ibid., p. 77.
11 Stocking, pp. 98–101. Haddon later wrote that he had 'no intention of

paying attention to ethnography, having been told by the experts in Britain that "there was little worth doing with regard to the natives" and had made no serious preparations'. Maybe so, but it seems possible that he had already decided to make a career move – away from marine zoology where he was overshadowed by Huxley and others – into a field where a young scientist could make a real name for himself.

12 J. Beete Jukes, on HMS *Fly* 1842–6, and John McGillivray, on HMS *Rattlesnake* 1846–50, were the main observers. Haddon reviewed the travel literature in TSR, Vol. I. J. Goodman, *The Rattlesnake. A Voyage of Discovery to the Coral Sea* (2005) is a modern account of one voyage.

13 J. Beckett, *Torres Straits Islanders: Custom and Colonialism* (Cambridge, 1987), pp. 24–60; TSR, Vol. I, pp. 3–18.

14 HP 1029; TSR, Vol. I, p. xi; Stocking, pp. 100–1; Quiggin, *Haddon*, pp. 81–90.

15 HP 1029.

16 HP 1029.

17 HP 1029. Haddon's attitude was always ambivalent. He identified strongly with the islanders, and mostly detested white traders and settlers, yet he also felt that 'progress' was inevitable.

18 HP 1029.

19 Quiggin, *Haddon*, pp. 91–2; R. Ackerman, *J. G. Frazer, His Life and Work* (Cambridge, 1987).

20 J. Urry, *Essays in the History of British Anthropology* (Chur, Switzerland, 1993); G. Jones, 'Contested territories: Alfred Cort Haddon, Progressive evolutionism and Ireland', *History of European Ideas* 24 (1998), pp. 195–211.

21 The Aran islanders continued to fascinate outsiders such as the Irish dramatist J. M. Synge, and the 'documentary' film-maker Robert Flaherty, who felt a constant need to romanticise and primitivise their way of life.

22 Haddon managed to be in two places at once by loading his teaching – at Cambridge in the spring; in Dublin in the autumn. He served on at least thirteen official committees.

23 J. Urry, 'Englishmen, Celts, and Iberians. The ethnographic survey of the United Kingdom 1892–1899', in *idem*, *Before Social Anthropology* (Chur, Switzerland, 1993).

24 Haddon, 'The saving of vanishing knowledge', *Nature* 55 (1897), pp. 305–6.

Chapter Two

1 Haddon to Geddes, 4 January 1897, HP 1048; Stocking, pp. 107–8.

2 *Man* (1939), 58–61. 'Ray himself delivered a paper in 1896 to the British Association stressing the urgent need for linguistic research in New Guinea. Ray also argued that sound ethnographic information could only be secured through a thorough knowledge of the language'. J. Urry, *Before Social Anthropology. Essays in the History of British Anthropology*, Chur, Switzerland, 1993, p. 75.

3 HP 1048; *ODNB* (Richard Davenport-Hines).

4 CSM/*HPA*; F. C. Bartlett, 'Charles Samuel Myers, 1873–1946', *ONFRS* (1948). Myers' father was Wolf Myers, his mother, Eugenie Moses.

5 Bartlett, *ONFRS*; Myers, CSM/*HPA*, p. 6.

6 HP 1048.

7 WMcD/*HPA*.

8 McDougall liked to make much of his Highland origins – his great-grandfather was a cobbler who had eloped across the border with an heiress and settled down in the north of England.

9 WMcD/*HPA*.

10 'An essay in which I foreshadowed the now fashionable doctrine of emergence of mind from the physical realm.' WMcD/*HPA*.

11 HP 1048.

12 Ibid.

13 Charles Seligmann changed his name to Seligman during the First World War. For simplicity's sake, he is here referred to as 'Seligman'. C. S. Myers, 'Charles Gabriel Seligman', *ONFRS* (1940), pp. 627–46.

14 HP 1048.

15 Ibid.

Chapter Three

1 H. Head, 'W. H. R. Rivers. An appreciation', *BMJ* (1922), 1, pp. 977–8; *idem*, 'William Halse Rivers Rivers, 1864–1922', *ONFRS* 95 (1923), pp. xliii–xlvii; A. C. Haddon (obituary of Rivers), *Man* 61 (1922), pp. 97–9; C. Myers, 'The influence of the late Dr Rivers on the development of psychology in Great Britain', in W. H. R. Rivers, *Psychology and Politics and Other Essays* (1923); C. G. Seligman, 'Dr W. H. R. Rivers', *Geographical Journal* 60 (1922), pp. 162–3. See also important obituaries by L. E. Shore and F. C. Bartlett in the St John's College magazine the *Eagle* 43 (1923), pp. 2–14.

2 D. Rockey, *Speech Disorder in Nineteenth-Century Britain. A History of Stuttering* (1980); G. W. Stocking, Jr., *Victorian Anthropology* (1987; New York, 1991), pp. 247–8, 251–2.

3 Quoted in Edward Wakeling, 'The real Lewis Carroll', talk to the Lewis Carroll Society, April 2003 (online access, May 2012). Rivers' sister Katherine later

recorded her memories of Carroll, but Wakeling questions their accuracy. M. N. Cohen and R. L. Green (eds.), *The Letters of Lewis Carroll* (New York, 1979).

4 Slobodin, p. 8.

5 P. N. Furbank, *E. M. Forster. A Life* (1977), pp. 40–4. Slobodin, p. 82. According to Grafton Elliot Smith, 'Rivers had always to fight against ill health: heart and blood vessels'; according to Seligman, he could only work for four hours a day. Slobodin, p. 9.

6 The atmosphere of Queen Square is best caught, not in the dull official history by Gordon Holmes but in the recollections of J. Purves Stewart, Ernest Jones, Macdonald Critchley, Foster Kennedy and other clinicians. Oliver Sacks has also made illuminating use of Hughlings Jackson's work.

7 *ODNB*, Myers.

8 The anaesthetist Joseph Lister was a cousin of Head. L. S. Jacyna, *Medicine and Modernism. A Biography of Sir Henry Head* (2008); C. Baumann, 'Henry Head in Ewald Hering's laboratory in Prague 1884–1886: An early study on the nervous control of breathing', *Journal of the History of the Neurosciences* 14 (2005), pp. 322–33.

9 Queen Square records. My thanks to Jackie Cheshire and the Medical Committee of the National Hospital for Nervous Diseases; J. Purves Stewart, *Sands of Time* (1939), pp. 39–40.

10 The Queen Square records contain printed copies of the testimonials provided in support of Rivers' job application in 1889 and summaries of the Board of Management discussions then and before his promotion to registrar in November 1890, but say nothing about the circumstances of his departure. Accounts of Rivers' life simply say that he 'resigned from' the hospital. However, the clinical notes he kept (which survive) show signs of declining performance. Rivers' interest in the problems of fatigue and nervous exhaustion, and their relationship to mental disorder, over the next few years are a further indicator that he had himself suffered a nervous breakdown – or come close to one.

11 Several sources (including the *ODNB*) incorrectly state that Rivers went to study with the physiologist Ewald Hering in Jena. In fact, Hering was still in Prague. Rivers' diary has disappeared. Myers and Shore quote from it. CSM/Influence.

12 *St Bartholomew's Hospital Reports* 29 (1893), p. 350; C. Gale and R. Howard, *Presumed Curable. An Illustrated Casebook of Victorian Psychiatric Patients in Bethlem Hospital* (Petersfield, 2003).

13 Bethlem Royal Archive. My thanks to Colin Gale. Photographs and brief case histories of some of Rivers' patients can be found in Gale and Howard, *Presumed Curable*.

14 *Under the Dome* (Bethlem Royal Hospital magazine), December 1892, March

1893. Thanks to Colin Gale for drawing this to my attention. H. Lee, *Virginia Woolf* (1996), p. 183.

15 C. Arnold, *Bedlam: London and its Mad* (2008); the standard history, J. Andrews et al., *The History of Bethlem* (1997), is more measured. M. Craig, *Psychological Medicine: A Manual on Mental Diseases for Practitioners and Students* (2nd edition, 1912) pp. 462–3.

16 *Mind* 4 (1895), pp. 400–3.

17 Rivers' 'German dream' in the First World War, WHR *C&D*, 165–80. May refer to a later visit, in 1896.

18 L. E. Shore, *Eagle* 43 (1923), pp. 2–12. In 1897 Rivers gave a lecture on 'fatigue' at Bart's and 'before he had finished, his title was writ large upon the faces of his audience. He had not yet acquired the art of expressing his ideas in an attractive form.' W. Langdon-Down, *Thus We Are Men* (1938), p. 30.; CSM/Influence; FCB/CE.

19 Slobodin, p. 17.

20 Shore, op. cit.; 'Rivers was a closeted homosexual', Langham, pp. 52 and 340n.7. The diary that Rivers kept while in Jena was found among his papers by Shore but has since disappeared. According to F. C. Bartlett, the difficulty of reading Rivers' handwriting was '*one* of the reasons why nobody has been terribly anxious to preserve these interesting writings' (italics added). Bartlett, British Psychological Society interview, 1959. For a careful assessment of the evidence relating to Rivers' sexuality, see JF/EF.

21 HP 1048.

22 Ibid.

23 Ibid.

24 Ibid.

Chapter Four

1 This and the following account of the voyage out is taken from Myers, 'Torres Straits Diary', Haddon's letters to his wife (HP 12/1) and Haddon's journal (HP 1,030).

2 G. W. Steevens, *Egypt in 1898* (1898), pp. 26–8.

3 Haddon to Fanny, HP 12/1.

4 Ibid.

5 HP 1048.

6 N. Stepan, *The Idea of Race in Science: Great Britain 1800–1960* (1982); G. Richards, '*Race', Racism and Psychology. Towards a Reflexive History* (1997).

7 T. H. Huxley, 'On the geographical distribution of the chief modifications on Mankind', *Journal of the Ethnological Society of London* 2 (1869–70), pp. 404–12.

8 D. Lorimer, 'Theoretical racism in late-Victorian anthropology, 1870–1900', *Victorian Studies* 31 (1988), pp. 405–30.

9 H. D. Rolleston, 'Description of the cerebral hemispheres of an adult Australian male', *Journal of the Royal Anthropological Institute*, 17 (1888), pp. 32–42.

10 H. Spencer, 'The comparative psychology of Man', *Popular Science Monthly* 8 (1876), pp. 257–69, reprinted in M. D. Biddiss (ed.), *Images of Race* (Leicester, 1979); E. G. Boring, 'The beginning and growth of measurement in psychology', *Isis* 52 (1961), pp. 238–57.

11 Urry, 'From zoology . . .', p. 75; Haddon, *Study*, pp. xxii, 18–19, 48–50; Urry, 'From zoology . . .', p. 65.

12 A. C. Haddon and A. H. Quiggin, *History of Anthropology* (1910), pp. 76–7; Stepan: 'Haddon denied that the Negro was closer, structurally, to the ape than whites. If anything the hairiness of the white brought whites closer, and the long legs and short body of the Negro brought Negroes further, from the evolutionary source. But "there can be no doubt that, on the whole the white race has progressed beyond the black race".' (p. 206).

Chapter Five

1 This account is based on Charles Seligman's diary (LSE), Haddon's journal (HP 1030) and letters to his wife (HP 12), Myers' journal (CUL Add MSS 8073) and Haddon's book *Head-Hunters*.

2 *Head-Hunters*, pp. 2–4; CGS/TSD; A. Herle and J. Philp, *Torres Straits Islanders An Exhibition Marking the Centenary of the 1898 Cambridge Anthropological Expedition* (Cambridge, 1998), p. 15.

3 CSM/TSD; Haddon to Fanny, 24 and 27 April 1898.

4 Wilkin to Browning, 5 July 1898, King's College, Cambridge.

5 *Head-Hunters*, p. 5; Slobodin, p. 23.

6 Ibid., pp. 8–10.

7 Ibid., p. 22.

8 Slobodin, p. 23, quoting Haddon's journal; J. Urry, 'Making sense of diversity', in Herle and Rouse.

9 CSM/TSD.

10 *Head-Hunters*, p. 23; Rivers, TSR, Vol. I, Pt I, p. i.

11 *Head-Hunters*, p. 37; Richards, 'Getting a result', in Herle and Rouse, pp. 139–44 – much the best account.

12 TSR, Vol. II, Part II, pp. 194, 201. According to Richards, McDougall was 'thus measuring flinching rather than pain thresholds, presumably on the highly dubious assumption that the former is an uncontrollable autonomic

response'. The algometer, designed by the American psychologist J. M. Cattell, was an instrument of experimental torture. A more sophisticated later version was known as the Dolorimeter.

13 R. M. Bache, 'Reaction time with reference to race', *Psychological Review* (1895), pp. 475–86; TSR, Vol. I, Part II, pp. 205–23. For auditory reaction times, Myers used an 'ad hoc instrument' designed for him – 'an electrically driven tuning fork, which inscribed a Deprez signal on the smoked surface of a hand-rotated drum'; for measuring visual reaction times, 'A large vertical screen, placed in the open air at a distance of about 35 metres from the subject and provided with a rectangular window, 20 × 5 centimetres. By electro-magnetic means a black board carrying a [white] card . . . was held up from behind the top of the screen. In this position the white surface was just hidden from view and the window appeared to the [subject] as black as the rest of the screen. When the board was released it fell through a certain distance, so that the window was now occupied by a white surface. The [subject] lifted his finger from a Morse key as soon as the white card became visible.'

14 S. Schaffer, *From Physics to Anthropology – and Back Again* (Cambridge, 1994), pp. 36–7.

15 C. S. Myers, 'The condition of life on a Torres Straits Island', *St Bartholomew's Hospital Reports* 35 (1899), pp. 91–9; CSM/TSD.

16 CSM/TSD.

17 Ibid.

18 Haddon says little about this period in *Head-Hunters*, but Seligman's diary records the party's progress around New Guinea.

19 Wilkin to Browning, 5 June 1898, King's College, Cambridge Archives. Seligman's diary is austerely scientific and seldom descends to personalities. Haddon may have been annoyed that, while other members of the expedition concentrated on their appointed tasks, Seligman extended his research into general anthropological areas that Haddon viewed primarily as his own.

20 Haddon to Fanny, 26 July 1898, HP 12/1.

21 Edith Myers interview with John C. Kenna, British Psychological Society Archive; WMcD/HPA; TSR, Vol. I, Part. II; CSM/TSD.

22 W. James, *The Principles of Psychology* (1890; 1901), Vol. I, pp. 192–3. James had once been a great fan of German experimental psychology, before losing interest and turning to philosophy.

23 *Head-Hunters*, p. 22.

24 Ibid., pp. 42–52.

25 CSM/TSD.

26 Ibid.; *Head-Hunters*, p. 47. Haddon photographed these scenes, but the 35 mm film cameras kept jamming and the footage that survives is very

disappointing. Haddon's will – and wilfulness – inspired a general revival of the old religion which had to be dealt with by the missionaries after he was gone.

Chapter Six

1 Haddon to Fanny, 20 July 1898, HP 12/1. This chapter is largely based on Myers' journal.
2 CSM/TSD.
3 Ibid.
4 B. Durrans, introduction to reprint of C. Hose, *Natural Man* (Singapore, 1988); R. H. W. Reece in *ODNB*; C. Hose and W. McDougall, *The Pagan Tribes of Borneo* (1912).
5 According to Haddon, Hose 'had promising young men to stay with him in his house for weeks at a time, and in this way they learn what a white man is like. Thus Mr Hose's residence is a sort of university whither the pupils come from all parts of his district to learn a little as to the meaning of government.' Quoted in B. Durrans, introduction to C. Hose, *Fifty Years of Romance and Research in Borneo* (1927; Kuala Lumpur, 1994).
6 CSM/TSD; Hose, *Fifty Years*, pp. 176–83.
7 WMcD/*HPA*.
8 CSM/TSD, 15 October 1898.
9 Ibid., 23 October 1898.
10 Ibid., p. 179. 'November 15, 16 or 17 I know not which', wrote Myers.
11 C. S. Myers, 'The beginnings of music', in *Essays Presented to William Ridgeway* (Cambridge, 1913).
12 Myers to Haddon, 16 January 1899, HP 1048.
13 HP 1048: Haddon to Fanny, 31 January 1899, HP 12/1. Haddon may have been irritated by a passage in Myers' letter which ran: 'I hope you found S to be better disposed than you thought he was when we parted. I cannot help thinking you are mistaken. I have known him a long time, but of course there is nothing like an expedition as a means for knowing a man.'

Chapter Seven

1 Scholars agree that Rivers got the idea of taking family histories from Francis Galton, 'whom he had consulted before leaving England' (Stocking, p. 112). Henrika Kuklick has suggested that Rivers may also have been influenced by the surveys of the late-Victorian poor conducted by social

researchers such as Booth and Rowntree which 'took the family or household as the basic unit of analysis'. Certainly in the medical research on colour vision which he had done prior to his trip to Torres Straits, Rivers had compiled family histories in order to assess whether deficiencies in vision were hereditary. So he was adapting medical skills. H. Kuklick, *The Savage Within. The Social History of British Anthropology 1885–1945* (Cambridge, 1991), pp. 140–1.

2 *Head-Hunters*, pp. 124–5.

3 W. H. R. Rivers, 'A genealogical method of collecting social and vital statistics', *Journal of the Royal Anthropological Institute* 30 (1900), pp. 74–82.

4 Haddon to Fanny, 19 September 1898, HP 12/1. A week later he added, 'I had a long talk with [Seligman] lately and things are now going more smoothly. S is very keen and hard working & I mean to stifle my personal dislike of him – for the cause of science – as he gains a lot of information that I have not time to collect.'

5 Langham, p. 69.

6 Haddon to Fanny, 25 September, 1 November 1898, HP 12/1. From the Grand Hotel, Thursday Island.

7 Haddon to Fanny, 5 April 1898, HP 12/1; Fanny to Haddon, 4 November 1898, HP 12/1.

8 Haddon to Fanny, 31 December 1898, 23 January 1899, HP 12/1.

9 Haddon to Fanny, 31 January 1899, HP 12/1.

10 Haddon to Fanny, 2 March 1899, HP 12/1; Haddon's journal.

11 Hose and McDougall, *Pagan Tribes*.

12 This account of the peace gathering is taken from Hose and McDougall, *Pagan Tribes*, Vol. I, pp. 289–300, an article by McDougall in the St John's College Magazine, the *Eagle*, quoted there, and Haddon, *Head-Hunters*, pp. 401–15.

Chapter Eight

1 Hose to Haddon, 3 August 1899, HP 1048/File 5.

2 Haddon to J. G. Frazer, 25 October 1899. Reprinted in Urry, *Before Social Anthropology*, pp. 79–81. Duckworth ended up as Master of Jesus College. Macalister, who had been both patron and treasurer of the expedition, seems to have been alienated by Haddon's casualness about money and habit of sending back anthropological artefacts at the university's expense. Many of these gathered dust and deteriorated while Haddon's future lay in the balance.

3 Quiggin, *Haddon*, pp. 110–19.

4 Haddon became a reader in 1909 and deputy curator of the museum in 1920. *ODNB*.

5 BAAS, Section H – Anthropology, Dover, 1899.

6 Other volumes on languages, arts and crafts, sociology, magic and religion followed over the decade, but Haddon did not get around to writing the general introductory volume, supposedly pulling the expedition's findings together, until 1935.

7 *Nature* 67 (1903), pp. 409–10; *TLS*, 1 May 1900. The anonymous reviewer was the policeman and retired colonial official, Basil Home Thompson.

8 Richards, 'Getting a result'.

9 TSR I, Vol. I, p. 45.

10 Rivers, 'Primitive Colour Vision', *Popular Science Monthly* 59 (1901), pp. 41–58. In 1858 Gladstone had called attention to the great vagueness of the colour terminology in the work of Homer and 'showed that Homer used terms for colour which indicated that his ideas of colour must have been different from our own, and he was inclined to go as far as to suppose that Homer had no idea of colour as we understand it, but distinguished little between differences of lightness and darkness'. This idea had then been taken further by other scholars but attacked on both scientific and literary grounds. Apart from anything, there was disagreement as to whether language is important in anthropology – whether language could be used 'as an indication of the mental development of a race'. By the end of the 1890s the idea of the evolution of a colour sense in man had been almost universally rejected. Rivers chose to revive it. 'We have progressive stages in the evolution of colour language: in the lowest there appears only to be definite terms for red and for white and black; in the next stage there are definite terms for red and yellow, and an indeterminate term for green; and a term for blue has been borrowed from another language; while in the highest stage there are terms of both green and blue, but these tend to be confused with each other. It is interesting to note that the order in which these four tribes are thus placed, on the ground of the development of their colour language, corresponds with the order in which they would be placed on the ground of their general intellectual and cultural development.'

11 WMcD/*HPA*.

12 A. C. Haddon, 'Appreciation', in E. E. Evans-Pritchard et al. (eds.), *Essays Presented to C. G. Seligman* (1934). Rivers, 'A genealogical method of collecting social and vital statistics', *Journal of the Royal Anthropological Institute* 30 (1900), pp. 74–82.

13 'I am resigned to the prospect of knocking around doing odd bits of anthropology and archaeology with a little journalism until I get married

and settled down,' he wrote to Oscar Browning in February 1900, after
publishing *Among the Berbers of Algeria*. King's College, Cambridge Archives.
14 HP 1048.

Chapter Nine

1 A. Moyal, *Platypus* (2002).
2 GES/Fragments; P. Morison, *J. T. Wilson and the Fraternity of Duckmaloi*
(Amsterdam, 1997); M. Macmillan, 'Evolution and the neurosciences down-
under', *Journal of the History of the Neurosciences* 18 (2009), pp. 150–96. Elliot
Smith's fellow hunters were his lifelong friends, J. T. Wilson and J. P. Hill.
3 G. Elliot Smith, 'The place of Thomas Henry Huxley in Anthropology',
Journal of the Royal Anthropological Institute 65 (1935), pp. 199–204.
4 GES/Fragments.
5 Elliot Smith to Wilson, 18 November 1896, and to Broom, 7 October 1896;
GES/Papers, BL Add MS 56303.
6 GES/Biog, p. 27; A. S. Eve, *Rutherford* (Cambridge, 1939), p. 26.
7 Elliot Smith to Wilson, 19 April 1900; GES/Papers, BL Add MS 56303;
Dart, 'Sir Grafton Elliot Smith', in Elkin and Mackintosh.
8 J. T. Wilson, 'Sir Grafton Elliot Smith', *ONFRS* 2 (1938), pp. 323–33; R.
Dart, 'Sir Grafton Elliot Smith and the evolution of man', in A. P. Elkin
and N. W. G. Mackintosh (eds.), *Grafton Elliot Smith: The Man and his
Work* (Sydney, 1974); A. Keith, *An Autobiography* (1950), p. 201; S. Zuckerman,
'Sir Grafton Elliot Smith', in S. Zuckerman (ed.), *The Concepts of Human
Evolution* (1973); Lord Rutherford, 'Early Days in Cambridge', in Dawson.
9 GES/Biog, pp. 30, 32.
10 G. Elliot Smith, 'The diversions of an anatomist', *Cambridge University
Medical School Magazine* 5 (1926) pp. 34–9.
11 GES/Biog, pp. 36–41; G. Elliot Smith, *The Royal Mummies. Catalogue générale
des antiquités égyptiennes du Musée du Caire* (Cairo, 1912; London, 2000).
12 D. M. Read, *Whose Pharaohs? Archaeology, Museums and Egyptian National
Identity from Napoleon to World War I* (Berkeley, 2002), p. 181.
13 M. A. Hoffman, *Egypt before the Pharaohs* (1980), pp. 249–64; J. A. Wilson,
Signs and Wonders upon Pharaoh (Chicago, 1964) pp. 144–58; Museum of
Fine Arts, Boston, Giza Archives Project: Reisner Biography (online access,
14 May 2011).
14 F. Wood Jones, 'In Egypt and Nubia', in Dawson. W. E. le Gros Clark,
'Frederic Wood Jones', *Biographical Memoirs of Fellows of The Royal Society*
1 (1955), pp. 119–34; H. A. Waldron, 'The study of the human remains
from Nubia: the contribution of Grafton Elliot Smith and his colleagues

to palaeopathology', *Medical History* 44 (2000), pp. 363–88 – an important article, scornful of Elliot Smith's and Wood Jones's work and of their reports on it.

15 Wood Jones, 'In Egypt and Nubia'.

16 Ibid.; A. J. E. Cave, 'A master anatomist', in Dawson, p. 195.

17 T. Wingate Todd, 'The scientific influence of Sir Grafton Elliot Smith', *American Anthropologist* 39 (1937), pp. 523–6.

18 Keith, *Autobiography*, p. 261; Certificate of Election and Candidature, Sir Grafton Elliot Smith, Royal Society Archive.

19 G. Elliot Smith, 'Some problems relating to the evolution of the brain', *Lancet* (1910), 1, pp. 1, 147, 221; *idem, The Evolution of Man* (1924; 1927). This book reprints lectures given in 1912, 1916 and 1924, with later material added in places – a dog's breakfast which makes it difficult to trace the evolution of Elliot Smith's ideas.

20 Elliot Smith, *Evolution of Man*, pp. 56–62.

21 Various theories about 'what went wrong with Elliot Smith, the brilliant anatomist?' can be found in S. Zuckerman (ed.), *The Concepts of Human Evolution* (1973). P. Crook, *Grafton Elliot Smith. Egyptology and the Diffusion of Culture* (Brighton, 2012) puts the defence case but lacks psychological insight. A. Stout, *Creating History. Druids, Ley Hunters and Archaeologists in Pre-War Britain* (Oxford, 2008) has two lively chapters on Elliot Smith and Perry.

22 *Nature*, 29 September 1910, p. 401; 13 October 1910, pp. 461–2; 20 October 1910, p. 495; 27 October 1910, pp. 529–30; 10 November 1910, p. 41; M. S. Drower, *Flinders Petrie. A Life in Archaeology* (1985) pp. 345–6.

23 W. H. R. Rivers, 'An address on "The Aims of Ethnology"', in WHRR/*P&P*, p. 116.

24 G. Elliot Smith, *The Ancient Egyptians and their Influence upon the Civilization of Europe* (1911); second edition, retitled *The Ancient Egyptians and the Origins of Civilization* (1923).

25 GES/Biog, pp. 51–3.

Chapter Ten

1 Head's 'most important contribution to the physiology of the nervous system [was his work on] the manner in which afferent impulses subserving sensation are integrated and conducted to the forebrain, and the function of the brain in integrating impulses of different nature and from various sensory organs to form the physiological basis of sensation. This has been one of the most obscure regions of neurology.' G. Holmes, 'Sir Henry Head', *BMFRS, ONFRS* 1941, p. 669.

2 L. S. Jacyna, *Medicine and Modernism: A Biography of Sir Henry Head* (2008), pp. 125–37, 144–5.

3 Ibid.

4 Ibid.

5 F. C. Bartlett, 'W. H. R. Rivers', *The Eagle* 62 (1968), pp. 156–60.

6 Jacyna, *Medicine, passim*. Henry and Ruth Head correspondence, Head papers, Wellcome Library. Ruth Head had as a young girl been photographed by Lewis Carroll, but when he sought permission to photograph her naked it was refused and his relations with her parents broken off. Jacyna, *Medicine*, p. 154.

7 W. H. R. Rivers, *The Influence of Alcohol and other Drugs on Fatigue* (1908).

8 W. H. R. Rivers, *The Todas* (1906).

9 'M. K. G.', *Bulletin of the American Geographical Society* 40 (1908), pp. 121–3.

10 R. L. Rooksby, 'W. H. R. Rivers and the Todas', *South Asia*, Vol. I (1971), pp. 109–22; Rivers, *The Todas*, p. 717, quoted in Stocking, p. 173.

11 Rooksby, 'Rivers and the Todas', pp. 114–15.

12 For reviews: *American Anthropologist* 9 (1907), pp. 196–7; *Man* 7 (1907), p. 258; *Folklore* 18 (1907), pp. 102–5. Frederic Bartlett admitted he never got through the book.

13 W. H. R. Rivers, *The History of Melanesian Society* (Cambridge, 1914), vol. I, p. 3.

14 Ibid., pp. 1–5; see also W. Sinker, *By Reef and Shoal. Being an Account of a Voyage amongst the Islands in the South-West Pacific* (1904) and C. Wilson, *The Wake of the Southern Cross. Work and Adventures in the Southern Seas* (1932).

15 Rivers, *History of Melanesian Society*, Vol. I, p. 2.

16 Langham, p. 125.

17 Hocart to Rivers in 1912, from Fiji, 'From what I hear . . . they are doing their best to spoil our Solomons. I met a Fijian in Suva who started talking Roviana; he told me my cook and brother was dead. A nice boy he was and a precocious dandy, and eager to come to Fiji. I seriously thought of fatiguing him to have flute duets with him and keep a memorial of heathen Roviana but the financial difficulty was great besides possible legal ones.' It isn't quite clear what 'fatiguing' means in this context. However, Rivers did not visit Roviana, so that any flute playing that went on would not have involved him. Langham, p. 125.

18 'Massage in Melanesia', 'Circumcision, incision, and subincision', 'Sexual relations and marriage in Eddystone Island of the Solomons', in WHRR/*P&E*.

19 WHRR/*P&E*, pp. 38–9.

20 W. H. R. Rivers, *Social Organization* (1924); WHRR/*P&P*, pp. 36–7.

21 FCB/CE.

22 Langham, p. 111 – unpublished typescript by Rivers, HP 12004.

23 GES/Biog, pp. 51–2.

24 GES/Biog, p. 56.

25 W. H. R. Rivers, 'The ethnological analysis of society', *Science* 334 (1911), pp. 385–97.

26 Rivers, 'Aims of Ethnology', in WHRR/*P&P* provides the basic diffusionist narrative. Rivers, 'The disappearance of useful arts', in WHRR/*P&E*.

27 Rooksby, 'Rivers and the Todas'.

28 G. Elliot Smith, *The Migrations of Early Culture* (Manchester, 1915) pp. 20–30.

Chapter Eleven

1 WMcD/*HPA*.

2 McDougall to Münsterberg, 18 September 1899. Münsterberg papers, Boston Public Library, MSS acc 1920 (2).

3 McDougall to Charles Taylor (Master of St John's), 31 May 1900; to R. F. Scott, 7 and 11 June 1900; to B. H. Farmer, 7 June 1900. St John's College Archives. McDougall's embarrassment caused him to write in so tortuous and convoluted a style that it took several letters before the council understood what he was saying. According to Boyd Hilton, McDougall could have remained as a tutor but he would have had to stay in college during the week. Thanks to Professor Hilton.

4 Wedding certificate copy, thanks to *ODNB* staff. Thanks also to Catherine Boyd for the census returns.

5 WMcD/Papers. Box 48.

6 WMcD/*HPA*.

7 WMcD/*HPA*.

8 Edith Myers interview with John C. Kenna, 1960, British Psychological Society Archive.

9 British Psychological Society papers, Wellcome Library; L. S. Hearnshaw, 'Sixty years of psychology', *Bulletin of the British Psychological Society*, January 1962, pp. 2–10; S. Lovie, 'Three steps to heaven. How the British Psychological Society attained its place in the sun', in G. C. Bunn et al. (eds.), *Psychology in Britain. Historical Essays and Personal Reflections* (Leicester, 2001).

10 W. H. R. Rivers, 'The colour vision of the natives of Upper Egypt', *Journal of the Royal Anthropological Institute* 31 (1901), p. 245; WMcD/*HPA*. For details of papers, A. L. Robinson (ed.), *William McDougall. A Bibliography* (Durham, NC, 1943).

11 Cyril Burt: 'This was part of an even larger problem that was occupying

him at that time, namely the desire to formulate a physiological basis for cognitive processes in terms of an organised hierarchy of nervous arcs, through which the unknown nervous energy ("neurin" as he called it) flowed in a pattern of waves guided by the resistance of the nerve-junctions or synapses. His first book *Physiological Psychology* was a remarkably compact and lucid survey of mental activity described from this standpoint. Written under the inspiration of Sherrington's recent work on "the integrative action of the nervous system", it sought to interpret Stout's cognitive psychology in terms of current neurological concepts and the dynamic psychology of James in terms of the Darwinian instincts, and to join the two into a coherent and consistent whole.' C. Burt, 'William McDougall: an appreciation', *British Journal of Educational Psychology* 9 (1939), pp. 1–7.

12 M. G. Brock, 'The Oxford of Raymond Asquith and Willie Elmhirst', in M. G. Brock and M. C. Curthoys (eds.), *The History of the University of Oxford. VII. Nineteenth-Century Oxford*, Pt 2 (Oxford, 2000).

13 R. Merrett, *A Jerseyman at Oxford* (1941); J. Howarth, '"Oxford for Arts": The Natural Sciences, 1880–1914', in Brock and Curthoys (eds.), *History of Oxford. VII*, Pt 2, p. 482. According to Howarth, 'the University disregarded [Wilde's] wishes and appointed the experimentalist William McDougall to the post'. She does not say why.

14 WMcD / *HPA*; J. A. Passmore, 'G. F. Stout's editorship of *Mind* (1892–1920)', *Mind* 85 (1976), pp. 17–36. To Anthony Quinton, McDougall's main scourges, John Cook Wilson and H. A. Prichard, 'were both slightly absurd figures: Cook Wilson, for his pugnacious eccentricity and long-windedness, Prichard for his leathery and obdurate dogmatism. They and their less talented ally, H. W. B. Joseph, took donnishness to its comical extreme, above all in the complacency with which they played Canute to the advances of mathematical and scientific knowledge.' A. Quinton, 'Thought', in S. Nowell-Smith (ed.), *Edwardian England* (Oxford, 1968). To Isaiah Berlin, Prichard 'was a very clever man. Totally dogmatic. He started from some very naïve premises, and then deduced, with great skill and ingenuity, various conclusions. The premises were dogmatic, but the methods of argument seemed fascinating and convincing.' B. Harrison (ed.) *Corpuscles* (Oxford, 1994).

15 Burt wrote of C. J. Flugel: 'For a Balliol man to desert Plato for Darwin and Spencer, attend lectures on physiology by Gotch and embark on psychological experiments in the Museum laboratory was something of a scandal. To study hypnotism was to incur a charge of dabbling in the black arts.' (Typescript obit, quoted in. A. Wooldridge, *Measuring the Mind. Education and Psychology in England c. 1860–c. 1990* (Cambridge, 1994), p. 57.) Burt, 'William McDougall: an appreciation'.

288 HEADHUNTERS

16 W. McDougall, 'The state of the brain during hypnosis', *Brain* 31 (1908). McDougall contributed articles on 'hallucination', 'hypnotism', 'subliminal self', 'suggestion' and 'trance' to the 11th edition of the *Encyclopaedia Britannica*, published in 1911.

17 S. Hynes, *The Edwardian Turn of Mind* (Princeton, 1968), pp. 132–8.

18 W. McDougall, *An Introduction to Social Psychology* (1908), p. 15: G. Richards, 'Defining a distinctively British psychology', *The Psychologist* 21 (July 2008); the *Introduction to Social Psychology*, says Leslie Hearnshaw, 'marked not only an epoch in McDougall's own development but an epoch in the history of psychology. It was a major influence in replacing the static, descriptive, purely analytical psychology of the nineteenth century with a psychology that was functional and dynamic.' L. Hearnshaw, *A Short History of British Psychology*, p. 188.

19 James W. Barclay, 'The race suicide scare', *Nineteenth Century*, December 1906, p. 895.

20 Mrs Alec Tweedie, 'Eugenics', *Fortnightly Review* 541 (1912), pp. 854, 861. These points are also made, without Mrs Tweedie's hysteria, in A. F. Tredgold, *Quarterly Review* 217 (1912), pp. 43–67.

21 W. McDougall, 'A practicable eugenic suggestion', *Sociological Papers*, III (London, 1907), pp. 55–104.

22 Irene M. Ashby MacFadyen, 'The birth rate and the mother', *Nineteenth Century*, March 1907, pp. 429–35.

23 W. McDougall, 'Psychology in the service of eugenics', *Eugenics Review* 5 (1914), pp. 295–308.

24 Hynes, *Edwardian Turn*, p. 139.

25 Quoted in J. Oppenheim, *The Other World. Spiritualism and Psychical Research in England, 1850–1914* (Cambridge, 1985), p. 393.

26 WMcD/HPA; R. van Over and L. Oteri (eds.), *William McDougall. Explorer of the Mind. Studies in Psychical Research* (New York, 1967); S. M. Mauskopf and M. R. McVaugh, *The Elusive Science. Origins of Experimental Psychical Research* (Baltimore, 1980), pp. 57–60.

27 WMcD/HPA.

28 W. McDougall, *Body and Mind* (New York, 1912; facsimile, 2006), pp. 347–54.

29 WMcD/HPA. It is unlikely that any of the Oxford philosophers read McDougall's books. The references in their papers attacking psychology, were mostly to psychologists of an earlier generation, such as G. F. Stout, McDougall's predecessor at Oxford, and James Ward.

30 B. Harrison, *Corpuscles*; Brock and Curthoys (eds.), *History of Oxford. VII*, Pt 2, p. 792.

Chapter Twelve

1 CSM/*HPA*; FCB/CE; FCB.

2 C. S. Myers, *A Textbook of Experimental Psychology* (1909); *idem*, *An Introduction to Experimental Psychology* (Cambridge, 1911).

3 'I imagine no psychologist who worked in those dismal rooms had the faintest idea as to the identity of St Tibbs', FCB/CE. Cambridge University also provided £50 for apparatus and expenses.

4 Myers to Münsterberg, 19 April 1912. Münsterberg papers, Boston Public Library.

5 C. S. Myers, 'Charles Gabriel Seligman', *ONFRS* 1941, p. 631; *idem*, 'The beginnings of music', in *Essays Presented to William Ridgeway* (Cambridge, 1913); *idem*, 'The ethnological study of music', in *Anthropological Essays presented to E. B. Tylor* (Oxford, 1907); M. Clayton, 'Ethnographic wax cylinders at the British Library Sound Archive: a brief history and description of the collection', *British Journal of Ethnomusicology* 5 (1996). For modern work see S. Mithen, *The Singing Neanderthals. The Origin of Music, Language, Mind and Body* (2005).

6 F. C. Bartlett, 'Charles Samuel Myers', *ONFRS*, p. 772.

7 B. McGuinness (ed.), *Wittgenstein in Cambridge. Letters and Documents, 1911–1951* (Malden, Mass., 2008), pp. 125–7; FCB/RDM.

8 T. H. Pear, 'Charles S. Myers', *British Journal of Educational Psychology* 17 (1947), pp. 1–5; *idem*, 'Charles Samuel Myers 1873–1946', *American Journal of Psychology* 60 (1947), pp. 289–96.

9 FCB/RDM.

10 C. Burt, 'Charles Samuel Myers', *Occupational Psychology* 21 (1947), pp. 1–5.

11 Ibid.

12 FCB/RDM.

13 F. C. Bartlett, 'Changing scene', *British Journal of Psychology* 57 (1956), pp. 81–7. No doubt living through two world wars had given Bartlett a new perspective on German psychology. In 1909 'Dr Rivers and his colleague Mr Myers showed some of the members a number of pieces of apparatus, including Kraepelin's modification of Musso's ergograph, a machine used in tests of memory, Stern's *Ton-variators*, Appunn's *TonMesser*, Runne's Chromoscope and a pendulum tachitoscope.' *Journal of Mental Science* 55 (1909), p. 392.

14 FCB/RDM; G. Thomson, *Eugenics Review* 38 (1947), pp. 187–8.

Chapter Thirteen

1 T. H. Pear, 'Personalities in the early days of the British Psychological Society', *British Journal of Psychology* 53 (1962), pp. 223–8.

2 B. Edgell, 'The British Psychological Society', *British Journal of Psychology* 37 (1947), pp. 113–31.

3 W. McDougall, 'An investigation of the colour sense of two infants', *British Journal of Psychology* 2 (1908), pp 338–52.

4 A. Wooldridge, *Measuring the Mind: Education and Psychology in England, c. 1860–c. 1990* (Cambridge, 1994); G. Sutherland, *Ability, Merit and Measurement* (Oxford, 1984); L. S. Hearnshaw, *Cyril Burt. Psychologist* (1979).

5 C.S. Myers, 'The pitfalls of mental tests', *British Medical Journal* (1911), i, pp. 1–8

6 S. Freud, 'Five lectures on Psycho-Analysis', in *idem, Two Short Accounts of Psycho-Analysis* (Harmondsworth, 1962).

7 H. F. Ellenberger, *The Discovery of the Unconscious. The History and Evolution of Dynamic Psychiatry* (New York, 1970) has a long chapter on Janet which was responsible for bringing him back into the public mind. B. Hart, *The Psychology of Insanity* (1912).

8 Freud, *Two Short Accounts of Psycho-Analysis*, pp. 44–5. Freud also pointed out that Janet's theory of hysteria reflected prevailing French views on heredity and degeneracy.

9 S. A. K. Wilson, 'Some modern French conceptions of hysteria', *Brain* 33 (1910–11), p. 298; S. Shamdasani, introduction to *Jung contra Freud* (Princeton, 2012).

10 Pear, 'Personalities'; W. Brown, 'Dreams: The latest views of Science', *Strand Magazine*, January 1913; T. H. Pear, 'The analysis of some personal dreams with reference to Freud's theory of dream interpretation', *British Journal of Psychology* 6 (1913), p. 303.

11 McDougall, *Body and Mind*, p. 327; *idem, Psychology. The Study of Behaviour* (1913), pp. 202–12; *idem*, 'The sex instinct', supplementary chapter to 8th edition of *An Introduction to Social Psychology* (1914).

12 *Proceedings of the Royal Society of Medicine* 7, Pt 3, Section of Psychiatry (1914).

13 Shamdasani, introduction to *Jung contra Freud*; Shamdasani argues that Freud's influence on Jung has been overstated: 'A Freudocentric legend arose, which viewed Freud and psychoanalysis as the principal source for all Jung's work. Jung himself said the "teachers that influenced me above all are Bleuler, Pierre Janet and Théodore Flournoy".' G. Makari, *Revolution in Mind. The Creation of Psychoanalysis* (2008), pp. 267–81.

14 Despite the vast literature on the Freud–Jung split, much remains unclear, most notably the role played by Jung's own mental health during this period.

15 S. Shamdasani, *Jung and the Making of Modern Psychology. The Dream of a Science* (Cambridge, 2003), pp. 219–30; WMcD/HPA.

16 R. A. Paskauskas, *The Complete Correspondence of Sigmund Freud and Ernest Jones 1908–1939* (Cambridge, Mass., 1993), p. 298; F. McLynn, *Carl Gustav Jung* (1996), pp. 226–7.

Chapter Fourteen

1 CSM/*HPA*.

2 For what follows CSM/*SSIF*, pp 1–14.

3 CSM/WD, 18 October 1914.

4 CSM/WD, 28–30 October 1914.

5 L. McDonald, *The Roses of No Man's Land* (1980; Harmondsworth, 1993), pp. 73–4.

6 CSM/CSS, 1. Privately Myers had his doubts about this patient. 'I was astonishingly successful for a preliminary canter, the man at once passing into light hypnosis – unless he is humbugging. I know him to be highly neurotic and a liar!' CSM/WD, 19 December 1914.

7 CSM/WD, 30 January 1915.

8 CSM/WD, 1–6 February 1915.

9 CSM/CSS, 1; CSM/WD, 27 February, 1915. The paper also described a third case seen by Myers.

10 Jones and Wessely; *WoN*; H. Binneveld, *From Shell Shock to Combat Stress. A Comparative History of Military Psychiatry* (Amsterdam, 1997).

11 M. I. Finucane, 'General nervous shock, immediate and remote, after gunshot and shell injuries in the South African campaign', *Lancet* (1900), 2, pp. 807–9.

12 A. G. Kay, 'Insanity in the army during peace and war, and its treatment', *Journal of the Royal Army Medical Corps* 18 (1912), pp. 146–58.

13 Equally strikingly, the military ignored the research on fatigue carried out by medical researchers, such as Rivers and McDougall in Britain and Kraepelin in Germany, before the war. A few military intellectuals such as J. F. C. Fuller referred to work on crowd psychology. *WoN*, p. 26.

14 CSM/*SSIF*, p. x.

15 Ibid.

16 G. Oram, *Military Executions during World War One* (Basingstoke, 2003), pp. 60–4; C. Corns and J. Hughes-Wilson, *Blindfold and Alone. British Military Executions in the Great War* (2001); CSM/*SSIF*, pp. 76–87.

17 CSM/*SSIF*, p. 90.

18 CSM/CSS, 1 and 2.

19 H. Cushing, *The Life of Sir William Osler* Vol. 2 (Oxford, 1925), p. 484.

20 W. A. Turner, 'Remarks on cases of nervous and mental shock', *BMJ* (1915), 1, pp. 833–5.

21 D. Forsyth, 'Functional nerve disease and the shock of battle', *Lancet* (1915), 2, pp. 1,399–403.

22 RWOCISS, pp. 4–6. Sir Alfred Keogh burned his papers, and the relevant War Office's files were largely destroyed by enemy bombing in 1940. To

resurrect official policy on shell shock, we are therefore obliged to rely on the post-war 'Report of the War Office Committee of Inquiry into Shell Shock' and odd remarks made by soldiers and doctors after the war. Gaps remain. So far as the public debate is concerned, newspaper coverage by and large remains unresearched.

23 Butler, p. 94. E. Bramwell, *Practitioner* 145, August 1940; H. C. Cameron, 'Sir Maurice Craig', *Guy's Hospital Reports* 85 (1935), p. 253.

24 CSM/*SSIF*, pp. 93–101.

25 Ibid., p. 95; RWOCISS, p. 145.

26 CSM/*SSIF*, pp. 95–6.

27 'Special discussion on shell shock', *Proceedings of the Royal Society of Medicine'*, (1916).

28 H. Wiltshire, 'A contribution to the aetiology of shell shock', *Lancet* (1916), 1, pp. 1,207–12. Dr Tracey Loughran has questioned whether any significance should be attached to Wiltshire's article. (T. L. Loughran, 'Shell shock in First World War Britain: an intellectual and medical history, c.1860–c.1920', PhD thesis, Queen Mary, University of London, 2005, p. 141.) Its importance can be overstated, but it is nonetheless indicative of a general shift in the shell shock literature. The fact that it was the first significant paper to be derived from work in France, rather than in England, remains valid.

29 IWM/Con[servation] Shelf. Arthur H. Hubbard, 'Letters written May–November 1916'.

30 W. Johnson and R. G. Rows, 'Neurasthenia and the war neuroses', in W. G. MacPherson, W. P. Herringham and T. R. Elliott (eds.), *History of the Great War. Diseases of the War* Vol. 2 (1923), pp. 1–67. This passage was written by William Johnson.

31 *WoN*, p. 41.

32 NA: WO95/695, War Diary of III Corps.

33 NA: WO95/45.

34 The Canadian history gave a lightly satirical account: 'The medical officer at the front had no knowledge of the jargon in which the problem was being discussed. He could not distinguish hypo-emotive from hyper-emotive, or commotion cerebri from emotio cerebri; he could not tell who was right about certain symptoms – Babinski, Claude, or Roussy, with their respective reflexes, dynamogenic, and dysocinetic explanations. "Rheumatism" he knew, a slacker he was pretty sure of after consultation with the sergeant major. All violent cases he classified in his own mind as "crazy", and sent them to a "special centre".' A. Macphail, *Official History of the Canadian Forces in the Great War 1914–19: The Medical Services* (Ottawa, 1925).

35 'Serious shock' was one suggestion. The Adjutant General preferred a twofold category – 'explosion (wound)' and 'nervousness'. By mid-November there was broad agreement that shell shock should be phased out and special treatment centres established.

36 NA: WO95/3980, War Diary of Director of Medical Services, Lines of Communication.

37 *WoN*, pp. 42–3; NA: WO95/3980, war diary of Sir Arthur Sloggett, director of medical services. On 15 August 1916, the DMS, L of C, 'forwards memo asking for our concurrence as regards Colonel Myers being known as Consulting Psychologist and only visiting bases the front [*sic*] when his services are asked for in consultation. Reply that we concur.' WO95/45. See also WO95/3980 – Myers was 'to visit bases in the front area when his services are asked for by Administrative MOs concerned'.

38 RWOCISS, pp. 123–5.

39 Information from Myers' daughter, Joan Rumens.

40 CSM/*SSIF*, pp. 84–5.

41 NA: WO 95/3980.

42 CSM/*SSIF*, p. 20.

43 These phrases, used by Holmes's Queen Square colleague, Foster Kennedy, reflect his thinking. Holmes said after the war that he had been 'impressed by the failures of the psychoanalysts and hypnotists'. As to shell-shocked soldiers, according to the neurologist Macdonald Critchley, 'he never liked those people'. *Brain* 130 (2007), pp. 288–93.44

44 Myers to Rivers, June 1917, Experimental Psychology Archives, Cambridge University.

45 Jones and Wessely, pp. 24–6; G. Roussy and J. Lhermitte, *The Psychoneuroses of War* (1918), pp. 164–5. Dr Maurice Chiray wrote in December 1916: 'We should not forget that most of our patients, and in particular post-traumatic reflex contractures, will almost always escape the neurological centres of the front. These subjects are evacuated as "wounded" and it is only in the interior that they become "nervous" and progressively acquire their functional deformity at the same time as they are being treated for their wounds.' Joseph Grasset added a further caution: 'it seems proven that too often [neurologists] are content merely to "whitewash" trauma victims and to send them back to the front incompletely cured'. Jones and Wessely have argued that the seminal influence which these experimental methods were to exercise not only on the French Army but also on other nations and subsequent conflicts was not based on the results it produced.

46 Butler, pp. 170–30. There is no adequate history of this stage in the shell shock story.

47 W. Brown, 'The treatment of cases of shell shock in an advanced neurological centre', *Lancet* (1918), 2, p. 197; F. Dillon, 'Treatment of neuroses in the field: the advanced psychiatric centre', in E. Miller (ed.), *The Neuroses in War* (1940); D. Carmalt Jones, 'War neurasthenia, acute and chronic', *Brain* 42 (1919), pp. 171–213; *idem, A Physician in Spite of Himself* (2009); W. Johnson and R. G. Rows, 'Neurasthenia and the war neuroses', in MacPherson et al. (eds.), *History of the Great War* Vol. 2, pp. 1–67.

48 Carmalt Jones, *Physician*, pp. 223–31. Carmalt Jones's real interest was in the physiological effects of shell shock.

49 E. Jones, A. Thomas and S. Ironside, 'Shell Shock: an outcome study of a First World War "PIE" unit', *Psychological Medicine* 37 (2007), pp. 215–24. The authors claim that the real statistics in the records for 4 Stationary Hospital at Arques were much less impressive than those publicly claimed: only seventeen per cent of the men admitted between 2 January and 9 November 1917 were returned to duty, while nineteen per cent were sent to other hospitals in France, thirty-five per cent were sent to convalescent camps, eight per cent were put on ambulance trains to the base and about twenty per cent were retained in base duties. As Carmalt Jones was much the most cautious of the doctors in making claims for his work, the implication is that the claims of seventy or eighty per cent success made by Brown and Dillon must be overblown. On relapses: at 12 GH in Rouen, Harold Wiltshire found that twenty-seven per cent of 150 cases of shell shock admitted had relapsed after an earlier breakdown. An inadequate survey by Gordon Holmes found that ten per cent of men relapsed once and 2.8 per cent relapsed more than once. Anecdotal evidence suggests the real rate was much higher. Brown, 'The treatment'. These figures matter because ever since the First World War, the rationale of military psychiatry – and the reason why armies have employed psychiatrists – has been its claim to be able, with early treatment, to send soldiers who have broken down back into the front line. PIE – proximity, immediacy, expectancy – is the backbone of modern military psychiatry in Iraq and Afghanistan.

50 Butler, pp. 127–8. In September 1918 the army finally abolished the classification of 'shell shock wound'.

51 CSM/*SSIF*, p. 22.

Chapter Fifteen

1 W. McDougall, *Janus, or the Conquest of War. A Psychological Inquiry* (1927), pp. 7–10. A strange, eloquent, impractical pacifist book.

2 P. Hoare, *Spike Island; The Memory of a Military Hospital* (2001); C. Stanford Read, *Military Psychiatry in Peace and War* (1920), pp. 40–5.

3 WMcD/*HPA*.

4 Ibid.

5 WMcD/*OAP*, pp. 218–19.

6 Ibid., pp. 272–3.

7 Ibid., p. 244.

8 Ibid., p. 237.

9 Ibid., pp. 10–17, 127–30, 534–7.

10 Ibid., p. 258. For the rich French neurological literature on fugues, see I. Hacking, *Mad Travelers. Reflections on the Reality of Transient Mental Illness* (Charlottesville, Virginia, 1998).

11 Loc. cit.

12 WMcD/*OAP* p. 267

13 Ibid., pp. 289–92.

14 A. Hurst, *Seale Hayne Neurological Studies* I (1918); *War Neuroses* (film), available from Pathé Archive; E. Jones, 'War Neuroses and Arthur Hurst', *Journal for the History of Medicine and Allied Sciences* 67 (2012). Percy Meek lived to seventy-five, did not marry, and died in King's Lynn in 1968.

15 H. Cushing, *From a Surgeon's Journal* (1936), p. 109. Almost every doctor who treated shell shock patients successfully immediately rushed into print, yet McDougall published almost nothing on shell shock during the war. 'During the war I wrote little but thought the more', was his later remark.

16 T. Saxby Good, 'The history and progress of Littlemore hospital', *Journal of Mental Science* 76 (1930), pp. 602–21; idem, 'The Oxford clinic', *Journal of Mental Science* 68 (1922), pp. 17–23.

17 T. W. Salmon, 'The care and treatment of mental diseases and war neuroses ("shell shock") in the British Army', originally published 1917. Reprinted in *The Medical Department of the United States Army in the World War*, Vol. X (Washington DC, 1929).

18 WMcD/*OAP*, pp. 266, 301, 248.

19 Good, 'The history and progress of Littlemore'. Wilfred Trotter's influential book, *The Instincts of the Herd in Peace and War*, had been published in 1916.

20 ibid.; idem, 'The Oxford clinic'; J. A. Hadfield, 'Psychological Medicine', *Medical Annual* (1921), pp. 373–88; idem, 'Suggestion and Hypno-analysis', in E. Miller (ed.), *The Neuroses in War* (London, 1940), pp. 128–49.

21 J. A. Hadfield, *Introduction to Psychotherapy* (1967), pp. 229–31.

22 W. Brown, C. S. Myers and W. McDougall, 'The revival of emotional memories and its therapeutic value', *British Journal of Medical Psychology* 1 (1920), pp. 16–33. See also C. G. Jung, 'The question of the therapeutic

value of abreaction', in *idem, Contributions to Analytical Psychology* (1928).
For an excellent discussion: R. Leys, *Trauma. A Genealogy* (Chicago, 2000).
23 Good, 'The Oxford clinic' and discussion.
24 W. McDougall, *National Welfare and National Decay* (1921), p. 29.

Chapter Sixteen

1 J. MacClancy, 'Unconventional character and disciplinary convention. John
Layard, Jungian and Anthropologist', *History of Anthropology* 4 (1986), pp.
50–71; *idem, Anthropology in the Public Arena: Historical and Contemporary
Contexts* (Chichester, 2013); J. Layard, 'Study of a failure', unpublished
autobiography, University of California, San Diego, Layard MSS; J. Layard
British Psychological Association interview, 3 May 1973.
2 Layard BPS i/v; autobiography.
3 Ibid.
4 The photograph is reproduced in Stocking, p. 201.
5 Layard BPS i/v.
6 MacClancy, 'Unconventional character'.
7 Elliot Smith to Perry, 1 and 5 February 1915. Perry papers, UCL; G. Elliot
Smith, *The Migrations of Early Culture* (Manchester, 1915); A. Stout, *Creating
Prehistory. Druids, Ley Hunters and Archaeologists in Pre-war Britain* (Oxford,
2008), pp. 82–7.
8 Elliot Smith, *Migrations*, p. 75; Elliot Smith to Haddon, March 1915. HP 3.
9 WHRR/*P&P*, p. 122; Elliot Smith to Perry, 23 April 1915. Perry papers,
UCL.
10 There was nothing resembling peer review. D. Forde, in Zuckerman (ed.),
Concepts of Human Evolution, p. 425; Elliot Smith to Perry, 17 May and 15
June 1915. Perry papers, UCL.
11 *Man*, November 1915.
12 WHRR/*P&E*, pp. 167–72.
13 B. Shephard, '"The early treatment of mental disorders": R. G. Rows and
Maghull 1914–1918', in H. Freeman and G. E. Berrios (eds.), *150 Years of
British Psychiatry. Volume II. The Aftermath* (1996).
14 Elliot Smith to Perry, 29 July 1915; Rivers to Perry, 11 and 18 August 1915.
Perry papers, UCL. Rivers addressed Perry as 'dear nephew' and often
ended his letters by sending greetings to Mrs Perry – 'Love to my niece'.
15 Shephard, 'Maghull'; G. Elliot Smith, 'Shock and the soldier', *Lancet* (1916),
1, pp. 853–7; G. Elliot Smith and T. H. Pear, *Shell Shock and its Lessons*
(Manchester, 1917), pp. 87–8.
16 Shephard, 'Maghull'. Among those trained to be a shell shock doctor at

Maghull was Charles Seligman, whose meticulous notes on the lecturing there survive in his papers at the London School of Economics.

17 Shephard, 'Maghull'; E. Jones, 'Shell shock at Maghull and the Maudsley; models of psychological medicine in the UK', *Journal of the History of Medicine and Allied Sciences* 65 (2010), pp. 368–95.

18 Layard BPS i/v; *idem*, autobiography, quoted in JF/EF.

19 Layard BPS i/v; *idem*, autobiography, quoted in JF/EF.

20 Shephard, 'Maghull'.

21 WHRR/*C&D*; p. 132.

22 WHRR/*C&D;* pp. 9–21.

23 'Freud's Psychology of the Unconscious', paper given to the Edinburgh Pathological Club, 7 March 1917. *Lancet*, 16 June 1917.

24 W. H. R. Rivers, 'The repression of war experience', reprinted in *idem*, *Instinct and the Unconscious* (Cambridge, 1922).

25 Layard, autobiography, quoted in JF/EF. (Jeremy MacClancy, *ODNB*.)

26 Sassoon papers, IWM; W. H. R. Rivers, 'Psycho-therapeutics', in J. Hastings (ed.), *Encyclopedia of Religion and Ethics* Vol. 10 (Edinburgh, 1918); Elliot Smith in WHR/*P&E*, p. xviii; H. Head, 'William Halse Rivers Rivers, 1864–1922', *Obituary Notices of Fellows Deceased. Proceedings of the Royal Society*, 95 (1923), pp. xliii–xlvii. D. L. Plowman (ed.), *Bridge into the Future. Letters of Max Plowman* (1944), p. 65. Plowman was a patient at Bowhill Hospital, a satellite of Craiglockhart. He had been 'knocked out' by a shell and invalided home suffering from concussion and shell shock.

27 *ODNB* (Rupert Hart-Davis; excellent). Biographies by John Wilson and Jean Moorcroft Wilson.

28 R. Hart-Davis (ed.), *Siegfried Sassoon. Diaries 1915–1918* (1983), p. 183; Sassoon Papers, IWM.

29 S. Sassoon, *Sherston's Progress* (1936; 2013), pp. 4–27; WHRR/*C&D*.

30 Sassoon, *Sherston's Progress*, p. 7.

31 Percy Smith to Rivers, July 1917. Experimental Psychology Archives, Cambridge University. Quoted in L. Stryker, 'Mental cases – British shell shock and the politics of interpretation', in G. Braybon (ed.), *Evidence, History and the Great War. Historians and the Impact of 1914–18* (New York, 2003); W. H. R. Rivers, *Instinct and Unconscious*(Cambridge, 1921), p. 186.

32 WHRR/*C&D*, pp. 156–7. He still saw some of his patients from Scotland and, indeed, began to feature in their dreams.

33 Sassoon papers, IWM.

34 W. H. R. Rivers, 'War Neurosis and military training', in W. H. R. Rivers, *Instinct and Unconscious* (Cambridge, 1921).

35 Head. 'He attempted to bring the abnormal phenomena of mental life into harmony with processes familiar on the physiological level. He conceived of

instincts as suppressed forms of primitive behaviour, which may be used in part through normal acts of consciousness or held in check completely. This he thought was analogous to the control exercised by "epicritic" impulses over more primitive "protopathic" reactions. Experience of the war neuroses led him to believe that, when a man regressed to a more instinctive form of conduct under the influence of mental states, his actions assumed an infantile character; this was evident in the content and structure of his dreams.'

36 Elliot Smith to Perry, 12 April, 18, 22, 24, 29 May, 6, 26 June 1915; 22 April, 6 October 1916. Perry papers, UCL.

37 Stout, *Creating Prehistory*, pp. 75, 82–7. For recent advocacy see P. Crook, *Grafton Elliot Smith, Egyptology and the Diffusion of Culture* (Brighton, 2012) and, wackier, J. D. Smith, *Egypt and the Origin of Civilization. The British School of Culture Diffusion* (Vindication Press, US, 2011).

38 Elliot Smith to Perry, 21 August 1916, Perry papers, UCL; Elliot Smith to Rivers, 4 November 1916, GES Papers.

39 Rivers, 'Aims of Ethnology', in WHRR/*P&P*.

Chapter Seventeen

1 J. Forrester, '1919: Psychology and Psychonanalysis. Cambridge and London – Myers, Jones and MacCurdy', *Psychoanalysis and History* 10 (2008), pp. 37–94.

2 CSM/*HPA*; CSM/*SSIF*, pp. 111–32.

3 Quoted in Forrester, '1919'.

4 C. S. Myers, 'Present-day Applications of Psychology, with special reference to Industry, Education and Nervous Breakdown' (1918), quoted in Forrester, '1919'. 'Rivers told us something of the new experimental psychology wh[ich] science he said had advanced more in the last 4 years than in the previous 100. Apparently our conception of mind has to be completely altered. A great deal of the work is due to the war . . . It is a new world wh[ich] looks as though it might have very wide influence on such matters as criminal law. What it is all based on is the operation of the unconscious mind wh[ich] profoundly affects the conscious & sometimes comes to the surface.' Alan Moore to Norman Moore, May 1919. C. Moore, *Hancox* (2010), p. 425.

5 Bunn et al., *Psychology in Britain*, pp. 100–3 and *passim*.

6 CSM/*HPA*; Forrester, '1919'.

7 FCB/RDM; CSM/*HPA*; Forrester, '1919'.

8 L. E. Shore, *The Eagle* 43 (1923), p. 11; P. Linehan (ed.), *St John's College Cambridge: A History* (Woodbridge, 2011), p. 453; A. Bennett, 'W. H. R. Rivers: Some recollections', *New Statesman*, 17 June 1922.

9 CSM/Influence; Shore, op. cit.

10 A. Bennett to Ellery Sedgwick, 29 December 1919, McMaster University Library, Special Collection; F. C. Bartlett, *The Eagle* 43 (1923), pp. 13–14; *Times Literary Supplement*, 5 May 1921. The review appeared anonymously. One of the mysteries of the shell shock saga is why Kinnear Wilson, the great Queen Square expert on French theories of hysteria, did not involve himself in the war. According to one hostile source, he spent the war 'bagging other men's practices'. I. K. Butterfield (ed.), *The Making of a Neurologist. The Letters of Foster Kennedy* (Hatfield, 1981).

11 JF/EF details relations between Rivers and the Freudians. It is clear from Ernest Jones's correspondence with Freud that he regarded Rivers as a figurehead to be used.

12 D. Rapp, 'The reception of Freud by the British press: General interest and literary magazines, 1920–1925', *JHBS* 24 (1988), pp. 191–202; Rivers to Jones, 27 November 1921; Jones to Rivers, 30 November 1921. Ernest Jones papers, British Psychoanalytical Society Archives.

13 Rivers, 'Aims of Ethnology', in WHRR/*P&P*; *idem*, *History and Ethnology* (1922).

14 Stocking, pp. 240–4; Langham, p. 147; Slobodin, p. 74.

15 F. C. Bartlett, *The Eagle* 43 (1923), pp. 12–14.

16 Shore, op. cit.

17 R. Hart-Davis (ed.), *Siegfried Sassoon. Diaries 1920–1922* (1981), pp. 166–8. Hours before the service was due to start his colleagues found Rivers' will dated 1920 and directing cremation, so plans were changed. T. R. Glover was unimpressed by the service: 'choir only full and some in ante chapel; lesson badly read by master, and rest Rootham's style of music. Poor Rivers!' Linehan (ed.), *History of St John's*, pp. 452–4.

18 Elliot Smith to Perry, 15 June 1922. Dawson, p. 79.

Chapter Eighteen

1 Elliot Smith now found himself at faculty meetings with his longstanding enemy, Flinders Petrie. The seating plan had to be redesigned to keep the two men as far apart as possible. M. S. Drower, *Flinders Petrie. A Life in Archaeology* (1985), p. 347.

2 F. Spencer, *The Piltdown Papers* (Oxford 1990) documents Elliot Smith's involvement.

3 R. Millar, *The Piltdown Men* (1974); Elliot Smith to Keith, 27 September 1913. Elliot Smith papers, John Rylands Library, University of Manchester; Keith, *Autobiography*, p. 327.

4 G. Elliot Smith, 'Primitive man', *Proceedings of the British Academy* (1915–16), p. 468; S. J. Gould, *The Panda's Thumb* (1983).

5 R. A. Dart, 'Australopithecus africanus: The Man-ape of South Africa', Nature 115 (1925), pp. 195–9; R. Lewin, Bones of Contention. Controversies in the Search for Human Origins (New York, 1987), pp. 50–2.

6 Langham, pp. 180–2.

7 Review of W. J. Perry, The Origin of Magic and Religion by T. E. Peet, Journal of Egyptian Archaeology 10 (1924), pp. 63–9.

8 M. Young, Malinowski. In June 1920, when Malinowski had just returned to England and had yet to publish his book, he met Rivers and Elliot Smith in Cambridge. Rivers was 'awfully nice' to him. 'Also I had a very long conversation with my arch enemy Elliot Smith in which he told me the long story of his conversion to his present ideas (which are on the whole phantastic) and then I passed some "psychological" criticism of his point of view. He told me that recently the same criticism or type of criticism has crossed his own mind ·& that he was shifting much more towards the psychological point of view. On the whole I liked Elliot Smith very much in a way: he has got at least a vigour of opinion & directness of views which are infinitely more congenial than the insipid lukewarmness of some of the others. [He] inquired whether I did any work about the daily life of the natives – the last question I would have expected from Elliot Smith!' Malinowski papers, 34/12, London School of Economics Archives.

9 Langham, p. 183.

10 Elliot Smith to Seligman, 9 June 1932. GES Papers.

11 S. Zuckerman, From Apes to Warlords (1978; 1988), p. 63.

12 Quoted in H. A. Waldron, 'The study of the human remains from Nubia: the contribution of Grafton Elliot Smith and his colleagues to palaeopathology', Medical History 44 (2000), pp. 363–88.

13 Quoted in ibid.

Chapter Nineteen

1 McDougall to Sherrington, 7 December 1920, Charles Sherrington papers, University of British Columbia.

2 W. McDougall, 'Is America Safe for Democracy?'

3 The enormous literature on intelligence testing in the US Army in the First World War can be found in L. Zenderland, Measuring Minds. Henry Goddard and the Origins of American Intelligence (Cambridge, 2001), p. 422.

4 In 1922, Walter Lippmann published six articles in the New Republic on 'The mental age of Americans', attacking the use of IQ tests, in which he lumped Stoddard and McDougall together. In The Great Gatsby, set in 1922 and published in 1925, Tom Buchanan, the husband of Daisy

Buchanan, the novel's principal woman character, alludes to Stoddard's book: "'Civilisation's going to pieces," broke out Tom violently. "I've gotten to be a terrible pessimist about things. Have you read 'The Rise of the Colored Empires' by this man Goddard?" "Why no," I answered, rather surprised by his tone. "Well, it's a fine book, and everybody ought to read it. The idea is if we don't look out the white race will be – will be utterly submerged. It's all scientific stuff; it's been proved.""

5 B. Kuklick, *The Rise of American Philosophy. Cambridge, Massachusetts, 1860–1920* (New Haven, 1977).

6 Ibid.

7 Ibid.; D. L. Kranz and D. Allen, 'The rise and fall of McDougall's instinct doctrine', *Journal of the History of the Behavioral Sciences* 3 (1967), pp. 326–38; H. Cravens and J. C. Burnham, 'Psychology and evolutionary naturalism in American thought, 1890–1940', *American Quarterly* 23 (1971), pp. 635–57. G. H. Sabine, *The Philosophical Review* 32 (1923), pp. 317–22; *American Historical Review* 26 (1921), pp. 748–9; *Journal of Philosophy* 18 (1921), pp. 690–7.

8 D. Cohen, *J. B. Watson: The Founder of Behaviorism* (1979).

9 J. B. Watson and W. McDougall, *The Battle of Behaviorism* (1928); H. Murray, autobiography, in E. G. Boring and G. Lindzey (eds.), *A History of Psychology in Autobiography*, Vol. 5 (New York, 1967). McDougall's doubts about the wisdom of applying behaviourism to child-rearing were to be borne out. Behaviourism became the dominant force in American academic psychology until the 1950s, inspiring such researchers as B. F. Skinner. In 1961 the British psychologist Donald Broadbent argued that behaviourism offered 'the best method for rational advance in psychology, allowing one to weed out facts from fantasy and to replace armchair speculation about the nature of the soul or the mysteries of consciousness with repeatable results'. By contrast Nehemiah Jorden doubted whether the behaviourists had made one positive contribution towards the increased knowledge of man. The historian David Cohen concludes that, while one 'needs to study human behaviour objectively', 'seventy years of research have shown that consciousness cannot be dismissed as uninteresting in human psychology and that even introspection has its uses as one tool among many.' D. Cohen, 'Behaviourism', in R. L. Gregory (ed.), *The Oxford Companion to the Mind* (Oxford, 1987).

10 S. H. Mauskopf and M. R. McVaugh, *The Elusive Science. Origins of Experimental Psychical Research* (Baltimore, 1980), pp. 17–24.

11 Ibid., pp. 20–1.

12 R. A Jones, 'Psychical, history and the press. The case of William McDougall and the *New York Times*', *American Psychologist* 42 (1987), pp. 931–40.

13 WMcD/HPA.

14 J. Bruner, autobiography, in G. Lindzey (ed.), *A History of Psychology in Autobiography*, Vol. 7 (San Francisco, 1980); Mauskopf and McVaugh, *Elusive Science* p. 83.

15 E. G. Boring, *A History of Experimental Psychology* (2nd edition, New York, 1950), p. 465; G. Allport in Boring and Lindzey (eds.), *A History of Psychology in Autobiography*, Vol. 5.

16 G. Murphy, *Historical Introduction to Modern Psychology* (new edition, 1967), p. 443; A. A. Roback, *History of American Psychology* (New York, 1952), pp. 257–9.

17 McDougall to Alexander, 18 May 1926. Samuel Alexander papers, John Rylands Library, Manchester.

18 R. F. Durden, *The Launching of Duke University, 1924–1949* (Durham, NC, 1993), p. 114.

19 McDougall to Alexander, 9 January 1932. Samuel Alexander papers, John Rylands Library, Manchester. To Sherrington he wrote: 'I am frankly funking the difficulties of the job of trying to do anything for English psychological medicine by direct action. I am afraid that I should spend my remaining years in fruitless efforts while I know that I have a clear line of work of a certain value before me, work which I can do pretty well.' McDougall to Sherrington, 22 December 1926. Sherrington ceased to be an active supporter.

20 McDougall to Gerald Heard, 17 December 1936. WMcD/Papers, Box 2, Folder 11.

21 Ibid., Box 2, Folders 11, 12, 14 and 15.

22 *The Spokesman-Review*, 8 August 1935; *New York Times*, 10 February 1935; Roback, *History of American Psychology*, p. 259.

23 WMcD/Papers, Folder 109; McDougall to Sherrington, 25 August 1925. Sherrington papers, University of British Columbia; Roback, *History of American Psychology*, p. 259. Abraham Roback had been a colleague at Harvard.

24 WMcD/HPA; Boring to Annie McDougall, WMcD/Papers, Folder 111.

25 W. McDougall, *Psycho-Analysis and Social Psychology* (1936), p. v.

26 J. C. Burnham, *Jelliffe* (Chicago, 1983), p. 236. Jung approved of his Lamarckian experiments; WMcD/HPA.

27 McDougall to Jones, 17 March 1925, Jones to McDougall, 11 May 1925. Ernest Jones papers, British Psychoanalytic Society. Freud to Jones, 20 May 1925. Paskausas (ed.), *Freud–Jung Correspondence*, p. 574.

28 Paskausas (ed.), *Freud–Jung Correspondence*, pp. 746–7; McDougall to E. V. Rieu (Methuen), 2 October 1935. WMcD/Papers, Folder 8.

29 J. S. Bruner, 'Autobiography' in G. Lindzey (ed.) *A History of Psychology in Autobiography*, Vol. 5 (New York, 1967), pp. 27–56.

30. F. A. E. Crew, 'A repetition of McDougall's Lamarckian experiment', *Journal of Genetics* 33 (1936), p. 61; McDougall's response appeared in the *British Journal of Psychology*. The rest of this chapter draws on WcMD/ Papers, Folder 48. 'Notes on his terminal illness. Written 6 October 1938.'
31 Ibid., Folders 16 and 17.
32 Myers to Annie McDougall, WMcD/Papers, Folder III; *Lancet* (1938), 2, p. 1387.

Chapter Twenty

1 A. Rodger, 'C. S. Myers in retrospect', *Bulletin of the British Psychological Society* 24 (1971), pp. 177–84. Other royal visitors included Queen Mary, who insisted on tackling a mechanical aptitude test, 'to the probably lasting detriment of her elbow-length white kid gloves', and ex-King Alphonso XIII of Spain, who gained no marks at all in a mechanical models test.
2 This account of the NIIP is largely taken from articles by Alec Rodger, Eric Farmer, C. B. Frisby and Winifred Raphael in the 1970 edition of *Occupational Psychology*.
3 H. J. Welch and C. S. Myers, *Ten Years of Industrial Psychology* (1932), p. 129.
4 W. Raphael, *Occupational Psychology* 44 (1970).
5 The relationship with Rowntrees was important. In 1922, Myers persuaded Seebohm Rowntree to appoint a full-time in-house psychologist at his chocolate factory in York. A decade later, Rowntrees, who were losing ground to their competitor Cadbury, commissioned an elaborate survey of its customers from the NIIP. The investigator, Nigel Balchin, found that 'chocolate assortments were primarily bought by men for women, a simple austere box was preferred; and that it might be a good idea to include a chart in the box identifying which chocolates have which centres'. The result was Rowntrees' Black Magic, launched in 1933 with a huge advertising campaign, which outsold Cadbury's Dairy Milk. This success could have led to further lucrative commissions but, with Cadbury's contributing £700 a year to the NIIP, it was decided that to prevent such conflicts of interest the Institute would in future avoid 'commercial' contracts and stick to purely neutral work. The final twist came in 1936 when Seebohm Rowntree was brought in to review the NIIP's performance and produced a report harshly critical of Myers' management style. G. Bunn, 'Charlie and the chocolate factory', *The Psychologist* 14 (2001), pp. 576–9.
6 Rodger, 'Myers in retrospect', *Occupational Psychology* 44 (1970).
7 Myers papers, Wellcome Library, PSY/MYE 1 and 2.

8   *Occupational Psychology* 44 (1970).

9   The membership was Lord Southborough (chairman), T. Beaton (Admiralty), J. L. Birley (Air Ministry), C. Hubert Bond (Board of Control), Sir Maurice Craig (Ministry of Pensions), Wing Commander Maurice Flack (Air Ministry), H. W. Kaye (Ministry of Pensions), Hamilton Marr (Board of Control for Scotland), Surgeon Captain E. T. Meagher (Admiralty), Colonel J. G. S. Mellor, Sir Frederick Mott, Major A. D. Stirling (War Office), W. Aldren Turner, Stephen Walsh MP, and Major W. Waring MP.

10 NA/WO32/4747.

11 *The Times*, 1 January 1940.

12 CSM/*SSIF*, p. ix.

13 *WoN*; R. H. Ahrenfeldt, *Psychiatry in the British Army in the Second World War* (1957).

14 *Occupational Psychology* 21 (1947), p. 15.

Conclusion

1   E. Lindstrum, 'The politics of psychology in the British Empire, 1898–1960', *Past and Present* 215 (2012), pp. 195–233.

2   A. C. Haddon, *Science* 34 (8 September, 1911), pp. 304–6; C. S. Myers, 'On the permanence of racial mental differences', in G. Spiller (ed.), *Papers on Inter-Racial Problems Communicated to the Universal Races Congress held at the University of London, July 26–29* (1911), pp. 73–9.

3   G.M. Frederickson, *Racism: A Short History* (Princeton, 2002), pp. 128–9.

4   R. S. Woodworth, 'Racial differences in mental traits', *Science* 31 (1910), pp. 171–86; B. Freire-Marreco and J. L. Myres (eds.), *Notes and Queries on Anthropology* (4th edition, 1912). Psychological tests of a different sort were later applied by British colonial administrators; see F. C. Bartlett, 'Psychological methods and anthropological problems', *Africa* 10 (1937), pp. 401–20; Lindstrum, 'Politics of psychology'.

5   E. B. Titchener, 'On ethnological tests of sensation and perception with special reference to tests of color vision and tactile discrimination described in the reports of the Cambridge anthropological expedition to Torres Straits', *Proceedings of the American Philosophical Society* 55 (1916), pp. 204–36.

6   Haddon, *Science* 34 (8 September 1911), pp. 304–6; Myers, 'Permanence of racial mental differences'.

7   E. Barkan, *The Retreat of Scientific Racism: Changing Concepts of Race in Britain and the United States between the World Wars* (Cambridge, 1992); N. Stepan, *The Idea of Race in Science: Great Britain, 1800–1960* (1982), pp. 140–69; G. Richards, '*Race*', *Racism and Psychology: Towards a Reflexive History* (1997).

8   C. S. Myers, 'Human Improvability', *Bristol Medico-Chirurgical Journal* 69

(1932), pp. 31–4. Rivers did warn that Lévy-Bruhl's idea of a 'prelogical mentality' might become a 'convenient label wherewith to label any manifestation of the human mind we do not understand'. Reflecting on his experience interviewing informants for hours at a time on arcane points of genealogy and ritual, he concluded that 'in intellectual concentration, as in many other intellectual processes, I have been able to detect no essential difference between Melanesian or Toda and those with whom I have been accustomed to mix in the life of our own society' (WHRR/PRE, pp. 53, 46). William McDougall 'approached the problem from the other direction, remarking on the power of religion, superstition and sentiment in every culture, and faulting Lévy-Bruhl "for the great error of assuming that the mental life of civilised man is conducted by each individual in a purely rational and logical manner"'. Lindstrum, 'Politics of psychology'.

9 Dawson, pp. 257–68.
10 Quoted in Barkan, *Retreat*, pp. 302–10.
11 Ibid.
12 'What makes the British?' *Oxford Today*, 23 July 2013. The researchers 'took samples from volunteers living in rural areas where all four grandparents had been born in the same area', and by this means 'obtained an accurate picture of the genetic make-up of rural Britain in around 1880, before the wide-scale population movements of the twentieth century or, more recently, immigration from other countries'.
13 H. Head, 'W. H. R. Rivers. An appreciation', *British Medical Journal* 1922 (i), pp. 977–8; *idem*, 'William Halse Rivers Rivers, 1864–1922', *Obituary Notices of Fellows Deceased. Proceedings of the Royal Society* 95 (1923), pp. xliii–xlvii; A.C. Haddon, *Man* 61 (1922), pp. 97–9; C. Myers, 'The influence of the late Dr Rivers on the development of psychology in Great Britain', *British Association for the Advancement of Science, Section J. Psychology* (1922), pp. 1–14, reprinted in W. H. R. Rivers, *Psychology and Politics and other Essays* (1923); C. G. Seligman, 'Dr. W. H. R. Rivers, *Geographical Journal* 60 (1922), pp. 162–3; G. Elliot Smith, 'Prefatory note' and 'A note on "The Aims of Ethnology"', in Rivers, *Psychology and Politics*; *idem*, 'Introduction: Dr Rivers and the new vision in ethnology', in W. H. R. Rivers, *Psychology and Ethnology* (1926); *idem*, 'Preface' in W. H. R. Rivers, *Conflict and Dream* (1923).
14 Stocking, p. 303; A. Kuper, *The Reinvention of Primitive Society. Transformations of a Myth* (Abingdon, 2005), p. 159.
15 A. R. Radcliffe-Brown, 'A further note on Ambrym', *Man* (March 1929), pp. 50–3; Kuper, *Reinvention of Primitive Society*, pp. 152-9; D. Forde in S. Zuckerman (ed.), *The Concepts of Human Evolution* (1973), p. 425.
16 E. Jones, *Free Associations* (1959), pp. 134–5; A. Compston, 'From the archives', *Brain* 134 (2011), pp. 920–3; F. M. R. Walshe, 'The anatomy and physiology

of cutaneous sensibility: a critical review', *Brain* 65 (1942), pp. 48–112.

17 E. Miller (ed.), *The Neuroses in War* (1940). Most of the contributors were on the staff of the Tavistock Clinic. They included the shell shock doctors Frederic Dillon, Millais Culpin and Arthur Hadfield.

18 Simon Schaffer interviewed by Alan MacFarlane, 27 June 2008. Internet access.

19 S. Sassoon, *Sherston's Progress* (1936), p. 67, quoted in Paul Fussell, *The Great War and Modern Memory* (1975, 1977), p. 101.

20 K. Rivers to Siegfried Sassoon, September 1936. Sassoon papers, IWM.

21 B. Shephard, 'Digging up the past,' *Times Literary Supplement*, 22 March 1996. The critic Bernard Bergonzi found 'something unconvincing about the whole work': B. Bergonzi, *War Poets and Other Subjects* (Farnham, 1999).

22 E. Jones and S. Wessely, '"Forward psychiatry" in the military: its origins and effectiveness', *Journal of Traumatic Stress* 16 (2003), pp. 411–19.

23 Childe quoted by Colin Renfrew in S. Zuckerman (ed.), *The Concept of Human Evolution* (1973), p. 438.

24 R. Lewin, *Bones of Contention* (New York, 1987).

25 F. Spencer, *The Piltdown Papers 1908–1955* (Oxford, 1990).

26 E. Leach in Zuckerman (ed.), *Concepts of Human Evolution*, p. 436.

27 D. Wilson, *Science and Archaeology* (1978); C. Renfrew, *Before Civilization: The Radiocarbon Revolution and Prehistoric Europe* (1973); S. Wells, *The Journey of Man: A Genetic Odyssey* (2002); A. Roberts, *The Incredible Human Journey: The Story of How We Colonised the Planet* (2009); B. Fagan, *Beyond the Blue Horizon: How the Earliest Mariners Unlocked the Secrets of the Oceans* (2012). Crook, *Grafton Elliot Smith* has a fair summary of Elliot Smith's current reputation.

28 W. McDougall, *An Introduction to Social Psychology* (University Paperback Edition, 1960), p. xxii.

29 S. Lentz, *Kenneth* (Bloomington, Indiana, 2007); R. Van Over and L. Oteri (eds.) *William McDougall: Explorer of the Mind. Studies in Psychical Research* (New York, 1967).

30 M. Boden, 'Purpose, personality, creativity', in G. C. Bunn et al. (eds.), *Psychology in Britain: Historical Essays and Personal Perspectives* (Leicester, 2001), p. 355.

31 R. Tallis, 'The neuroscience delusion', *Times Literary Supplement*, 9 April 2008; R. Tallis, *Neuromania, Darwinitus and the Misrepresentation of Humanity* (Durham, 2011). On what neuroscience can and cannot do: S. Satel and S. O. Lilienfeld, *Brainwashed: The Seductive Appeal of Mindless Neuroscience* (New York, 2013); C. Renfrew et al. (eds.), *The Sapient Mind: Archaeology Meets Neuroscience* (Oxford, 2009).

32 R. Tallis, 'Neurotrash', 10 November 2009. Internet access.

33 Quoted in S. Zuckerman (ed.), *The Concept of Human Evolution* (1973), p.14.

Index

www.vintage-books.co.uk